Bookware Companion Series™

Simulations of Machines
Using MATLAB® and Simulink®

Books in the BookWare Companion Series™

Bookware Companion Series™

Simulations of Machines
Using MATLAB® and Simulink®

John F. Gardner
Boise State University

BROOKS/COLE

THOMSON LEARNING

Australia • Canada • Mexico • Singapore • Spain • United Kingdom • United States

BROOKS/COLE

THOMSON LEARNING

Sponsoring Editor: *Bill Stenquist*	Permissions Editor: *Sue Ewing*
Marketing Team: *Chris Kelly, Ericka Thompson, Carla Martin-Falcone*	Cover Design: *Denise Davidson*
	Cover Illustration: *Craig Borghesani*
Editorial Coordinator: *Shelley Gesicki*	Print Buyer: *Nancy Panziera*
Production Editor: *Laurel Jackson*	Typesetting: *Scratchgravel Publishing Services*
Production Service: *Scratchgravel Publishing Services*	Printing and Binding: *Webcom, Ltd.*

COPYRIGHT © 2001 Wadsworth Group.
Brooks/Cole is an imprint of the Wadsworth Group, a division of Thomson Learning, Inc.
Thomson Learning™ is a trademark used herein under license.

For more information about this or any other Brooks/Cole product, contact:
BROOKS/COLE
511 Forest Lodge Road
Pacific Grove, CA 93950 USA
www.brookscole.com
1-800-423-0563 (Thomson Learning Academic Resource Center)

ALL RIGHTS RESERVED. No part of this work covered by the copyright hereon may be reproduced or used in any form or by any means—graphic, electronic, or mechanical, including photocopying, recording, taping, Web distribution, or information storage and retrieval systems—without the prior written permission of the publisher.

For permission to use material from this work, contact us by
www.thomsonrights.com
fax: 1-800-730-2215
phone: 1-800-730-2214

MATLAB and Simulink are registered trademarks of The MathWorks, Inc. Further information about MATLAB and related publications may be obtained from:

The MathWorks, Inc.
3 Apple Hill Drive
Natick, MA 01760
(508) 647-7000; Fax: (508) 647-7001
e-mail: info@mathworks.com
www.mathworks.com

Printed in Canada

10   9   8   7   6   5   4   3   2   1

Library of Congress Cataloging-in-Publication Data
Gardner, John F. (John Francis), [date]–
   Simulations of machines using MATLAB & SIMULINK / John F. Gardner.
      p. cm. — (BookWare companion series)
   Includes index.
   ISBN 0-534-95279-8
      I. Title.   II. BookWare companion series (Pacific Grove, Calif.)

TJ153 .G27   20001

00-048713

To my father, Richard,
who taught me my first lessons in engineering;
and to my mother, Mary Elizabeth,
who taught me lasting lessons about strength

About the Series

"The purpose of computing is insight, not numbers."
—R.W. Hamming, *Numerical Methods for Engineers and Scientists,* McGraw-Hill, Inc.

It is with this spirit in mind that we present the BookWare Companion Series.™

Increasingly, the latest technologies and modern methods are crammed into courses already dense with important theory. The question is asked: "Are we simply teaching students the latest technology, or are we teaching them to reason?" We believe these two alternatives need not be mutually exclusive. This series was founded on the belief that computer solutions and theory can be mutually reinforcing. Properly applied, computing can illuminate theory and help students to think, analyze, and reason in meaningful ways. It can also help them understand the relationships and connections between new information and existing knowledge; and cultivate in them problem-solving skills, intuition, and critical thinking. The BookWare Companion Series was developed in response to this mission.

Specifically, the series is designed for educators who want to integrate their curriculum with computer-based learning tools, as well as for students who wish to go further than their textbook alone allows. The former will find in the series the means by which to use powerful software tools to support their course activities, without having to customize the applications themselves. The latter will find relevant problems and examples quickly and easily and will have electronic access to them. Important for both educators and students is the premise upon which the series is based—that students learn best when they are actively involved in their own learning. This series will engage them, provide a taste of real-life issues, demonstrate clear techniques for solving real problems, and challenge students to understand and apply these techniques on their own. The books in the series all encourage active learning.

To serve your needs better, we plan to improve the series continually. Join us at our BookWare Companion Resource Center web site (*http://www.brookscole.com/ engineering/ee/bookware.htm*) and let us know how to improve the series; share your ideas on using technology in the classroom with your colleagues; suggest a great problem or example for the next edition; let us know what is on your mind. We look forward to hearing from you, and we thank you for your continuing support.

Bill Stenquist	Publisher	*bill.stenquist@brookscole.com*
Shelley Gesicki	Editorial Coordinator	*shelley.gesicki@brookscole.com*
Chris Kelly	Marketing Manager	*christopher.kelly@brookscole.com*
Ericka Thompson	Marketing Assistant	*ericka.thompson@brookscole.com*

CONTENTS

Cʜᴀᴘᴛᴇʀ 4

Kɪɴᴇᴍᴀᴛɪᴄ Sɪᴍᴜʟᴀᴛɪᴏɴs Usɪɴɢ Sɪᴍᴜʟɪɴᴋ   28

Cʜᴀᴘᴛᴇʀ 5

Iɴᴛʀᴏᴅᴜᴄɪɴɢ Dʏɴᴀᴍɪᴄs   53

CHAPTER 9

THE TREBUCHET   98

APPENDIX

SIMULINK TUTORIAL   109

PREFACE

This book is intended primarily for use as a supplemental textbook for studying mechanisms at the undergraduate level. However, it also can be a starting point for a senior-level elective in dynamics, as we have done at Penn State. The ideas for the book originated in discussions among my colleagues about how dynamics is introduced in a typical undergraduate kinematics class. Traditionally, the study of kinematics has been highly structured: Students are led through a series of steps that include position, velocity, and acceleration solutions. Only after detailed motion analyses are students introduced to the concept of forces and, hence, dynamics. The apparent contradiction—analyzing motion before forces—results from the constrained nature of mechanisms and from certain, often unstated assumptions that some ideal prime mover exists that can generate whatever torque/force is needed to maintain a constant speed of one of the links. The concept of true dynamic simulations of constrained mechanisms is considered outside the scope of traditional undergraduate kinematics. However, the reasons for these attitudes are historical and somewhat outdated.

In this book, we take advantage of a family of dynamics-oriented computer packages typified by MATLAB® and its add-in, Simulink®. By combining the concepts of vector loop equations (to express kinematic constraints), numerical simulation (to compute velocities and positions when accelerations are known), and matrix algebra (to compute accelerations and constraint forces simultaneously), this book guides students through the steps required to assemble a fully functional dynamic simulation of a mechanism. This type of simulation is an essential design tool because it gives engineers access to internal forces needed in sizing bearings and members.

Although instructors can use this book as a standalone text or as a reference for practical engineers, it is primarily intended as a supplemental text in undergraduate mechanisms classes. The chapters are presented in logical order, but their order does not depend strongly on material in previous chapters. For instance, Chapter 2 offers a review of vector loop equations and their derivatives. If students are already competent in this area, then it can safely be skipped. Likewise, Chapter 3 describes the use of MATLAB to solve the nonlinear position problem in kinematics; this chapter could also be skipped because the solution of this problem is performed implicitly by the simulation package. The table in this preface is a guide that shows how instructors can use the chapters in this book to supplement other textbooks in the field.

Cross Reference of Chapters in This Book with Traditional Kinematics Textbooks

Chapter in This Book	Shickley & Uicker	Erdman & Sandor	Mabie & Reinholtz	Norton	Waldron & Kinzel
2	2, 3	3	2	2, 4	3
3	2	3	2	4	2, 3
4	3, 4	3, 4	8	6, 7	2
6	14, 15, 16	5	9	10, 11	11, 12
Case Studies	5, 17, 18	6, 8	3, 11, 12	8, 9, 13, 15	

While I wrote this book, MATLAB and Simulink went through several revisions. I began the book with MATLAB 4.3 and Simulink 2.0, and by the time I finished it, I was using R11.1 of the MATLAB/Simulink package. I have attempted to make the files backward-compatible, but many of the revisions during that time have affected the integration algorithms that lie at the core of Simulink. Therefore, for best performance of the simulations, I recommend that you use the most recent release.

Acknowledgments

I would like to acknowledge the many friends and colleagues—more than I can possibly enumerate here—from whom I have received much help and support. I want to thank John Lamancusa, Joe Sommer, and Marty Trethewey; their comments and perspectives led me to the ideas for this book. Steve Velinsky, of the University of California at Davis, was host to my sabbatical leave, during which time much of this book was developed. Jonathan Plant and Bill Stenquist, my editors at Brooks/Cole (and its forerunner, PWS), encouraged me and prodded me along on this project. Bill Murray at the California Polytechnic State University, San Luis Obispo, piloted early drafts of this book in his classes, and he and his students provided excellent feedback. Stephen Derby at Rensselaer Polytechnic Institute, Imme Ebert-Uphoff at the Georgia Institute of Technology, Greg Luecke at Iowa State University, Mark Nagurka at Marquette University, and William J. Palm at the University of Rhode Island also provided comments and suggestions. I also want to thank Dina Berkhoff, Matt Lichter, Jenny Rincon, Will Reutzel, and Gayathri Vijayakumar, very patient students who were subjected to early versions of this book; the final version is much improved for all their comments. Naomi Bullock and Jane Carlucci at The MathWorks, Inc., provided technical support.

Finally, no list of acknowledgments would be complete without mention of my wife, Barbara Bowling, and my daughters, Sarah and Beth. Their support, love, and patience are a constant source of strength for me.

John F. Gardner
Boise, Idaho

Bookware Companion Series™

Simulations of Machines
Using MATLAB® and Simulink®

CHAPTER 1

INTRODUCTION AND OVERVIEW

1.1 Why Simulate Mechanisms?

The term *computer simulation* refers to a technique in which mathematical models are developed to capture some important characteristics of a system under study. These models, often in the form of time-dependent ordinary differential equations, are then solved numerically in a computer simulation. Simulations are used to study a wide range of systems from internal combustion engines to the stock market. The purpose of this book is to investigate the use of computer simulations to analyze mechanisms.

The purpose of a mechanism is, in general, to provide some predefined motion over time. The fundamental fact that mechanisms are intended to move suggests that computer simulations would be useful in studying them. However, textbooks in kinematics have been reluctant to embrace computer simulations because formulation of the equations of motion require, in general, graduate level understanding of dynamics (e.g. Lagrangian or Hamiltonian mechanics). This text outlines two approaches that can be used to allow students to use existing computer simulation packages to help solve kinematic equations (kinematic simulations) and to formulate complete dynamic simulations using Newtonian methods that they have already learned in sophomore-level mechanics classes.

In this text, computer simulation tools will be introduced on two levels. First, numerical integration, the core technology of computer simulations, will be used to bypass the position problem, which is arguably the most difficult portion of motion analysis (which consists of position, velocity, and acceleration analyses). Students are shown how to use Simulink to analyze a mechanism that is moving with both constant and varying input speeds, with and without acceleration analyses. Later, these ideas are expanded to include the notion of dynamics. Using a technique that is best known from textbooks by Haug,[1] the vector loop equations are combined with Newton-Euler formulations of dynamics to provide a linear set of equations. These can be assembled in matrix form and the simulation can compute both accelerations and forces for each time step of the simulation. An added benefit of this method is that it produces more motion

---

[1] Haug, E. J., *Intermediate Dynamics,* Englewood Cliffs, N.J.: Prentice Hall, 1992.

coordinates than required to uniquely define the mechanism position. A test of consistency among the coordinates serves to monitor errors both in the problem formulation (debugging) and numerical integration (simulation stability).

1.2 Kinematic Simulations

The term *kinematic simulation* refers to the use of computer simulation packages to solve the kinematic equations of a mechanism over time, thus providing the velocities and accelerations of various links through continuous motion of the mechanism. Kinematic simulations make use of the geometric properties of the mechanism (i.e. link lengths). Figure 1-1 provides an overview of the kinematic simulation as it might be implemented in a block diagram-oriented simulation package such as MATLAB/ Simulink.

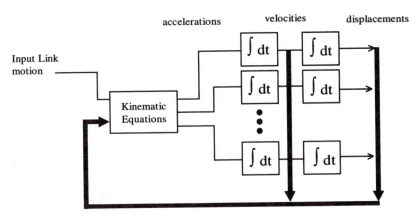

FIGURE 1-1   Block diagram representation of the kinematic simulation method.

In Figure 1-1, the box labeled "Kinematic Equations" represents a user-defined program block in which the equations that represent the relationships between the accelerations of the links are solved. Chapter 2 reviews vector loop equations that will be fundamental in formulating those equations.

There are important differences between this approach to kinematics and computer implementations of the traditional approach. In this case, the numerical integration algorithms that are built into the simulation package are used to compute velocities and displacements from the accelerations. In other words, only the acceleration analysis need be accomplished and embodied in the function block. The numerical integration will provide all the velocity and displacement information required. Chapter 4 will discuss this technique in greater detail and show its implementation for a slider-crank mechanism and a four-bar mechanism.

1.3 Dynamic Simulation of Mechanisms

As discussed previously, full dynamic simulation of mechanisms have historically been considered outside the scope of undergraduate kinematics courses. Indeed, the traditional approach using Lagrangian methods for constrained systems is appropriately taught at the graduate level. On the other hand, undergraduate students are well-versed in Newtonian mechanics applied to rigid bodies. Therefore, it seems reasonable to formulate a dynamic simulation by applying Newtonian methods to each individual link and combine the resultant force/acceleration relationships with the acceleration relationships that reflect the constraints of the mechanism itself. These equations constitute a set of simultaneous algebraic equations that are linear in the link accelerations (translational and rotational) and constraint forces. Therefore, these equations can be compiled in matrix form and the simulation package can solve for the forces and accelerations at each time step.

Note that there is a fundamental difference between this approach and the traditional one of Lagrangian or Hamiltonian mechanics. In the classical approaches, the equations of motion are reduced to the lowest order. In other words, a four-bar linkage, since it has only one degree of freedom, can be represented by a single second-order differential equation. On the other hand, the computer simulation outlined in this text will be solving a second-order differential equation for each link, the solutions of which are not independent. In fact, as will be shown in Chapters 4 and 6, this fact is used to provide external checks on the validity and self-consistency of the simulation.

The computer simulation for this system can be represented by the generic block diagram shown in Figure 1-2.

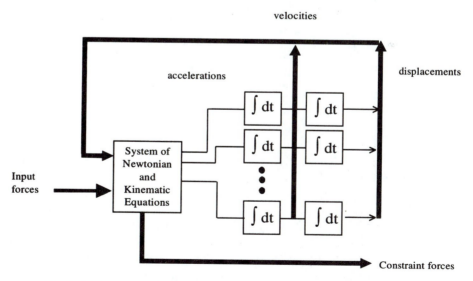

FIGURE 1-2 Generic block diagram showing structure of mechanism simulation using the simultaneous constraint method.

1.4 Summary

This book presents an innovative method of analyzing and designing constrained mechanical systems or mechanisms. While the book is focused on the use of MATLAB and Simulink to carry out the resulting simulations, the theoretical foundations presented in Chapters 2, 3, 4, and 6 are not unique to the computer method used to solve the equations.

CHAPTER 2

VECTOR LOOP AND VECTOR CHAIN EQUATIONS

2.1 Introduction

At the core of nearly all attempts to automate kinematic analyses lies the vector loop equation. The vector loop equation is a very compact and elegant expression of the constraints that exist between the various bodies of a mechanism. Vector loop equations are easy to formulate and are the necessary first step in computerized analysis of mechanisms.

Similarly, vector chains can be used to indicate the position of a point, such as a center of mass of a link, which is important for the analysis, but not at one of the mechanism joints. Vector chains are also used to analyze mechanisms that are open-chain—that is, those that have only one joint with the fixed link or ground.

The methods summarized here are very similar to techniques found elsewhere. Kinematic analysis with complex numbers as well as "rotating vectors" are closely related and, indeed, provide the same mathematical relationships as vector loop equations. It is the opinion of the author that this presentation is more accessible than the other methods.

2.2 The Planar Vector

A vector, as all undergraduate engineering students know, is a mathematical concept used to represent physical quantities which have both magnitude and direction. Simply stated, a displacement vector represents a directed distance between two points in space. For mechanism analysis, it can be seen that each link in a mechanism can be represented as a displacement vector, the tail of which is at the center of one joint and the head at the center of the other. The magnitude of this displacement vector is the link length while the angle the base of the vector makes with the positive x-direction is the angle of the link (counterclockwise positive). Figure 2-1 illustrates these points.

Note that the vector is represented in bold, uppercase notation, $\mathbf{R}$, while the length of the vector, r, is in lowercase and not boldface. This notation will be used consistently throughout the book.

Given the coordinate frame shown in Figure 2-1, the x- and y-components of the vector $\mathbf{R}$ can easily be written in terms of the vector length r and the angle with respect to the positive x-axis; $r_x = r \cos(\theta)$ and $r_y = r \sin(\theta)$.

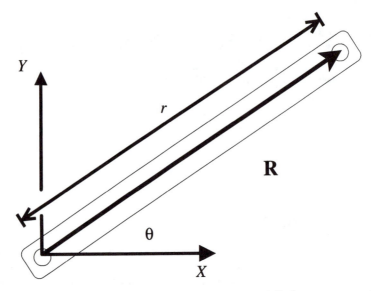

FIGURE 2-1   Single link of length r with coordinate frame and displacement vector, **R,** shown.

2.3   Single Loop Equations

Now the concept of the vector loop equation (or loop closure equation) is presented. Figure 2-2 shows a four-bar mechanism that is a common example in kinematic texts. The numbering system adopted here numbers the ground link as 1, so the movable links start at number 2.

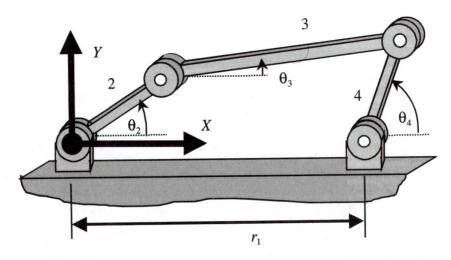

FIGURE 2-2   Four-bar mechanism with x-y coordinate frame attached at the leftmost base pivot. Note that the x-axis lies along the line that passes through both base pivots.

As seen in Figure 2-2, a coordinate system is assigned to the system. Convention holds that a coordinate system is placed at the ground joint, which is leftmost in the drawing, although this is completely arbitrary. The coordinate system can be placed anywhere as long as subsequent operations are performed in a manner that is consistent with it.

Each link (including the ground link) is represented by a displacement vector, as described in the previous section. Figure 2-3 shows the same four-bar mechanism with four displacement vectors drawn in place.

A few comments are in order concerning the placement of the vectors. As with the coordinate system, there are a number of ways in which these vectors may be drawn. As we will see, any set of displacement vectors will lead to a valid vector loop equation as long as the equation is derived in a manner that is consistent with the vectors.

Note that vectors $\mathbf{R}_2$ and $\mathbf{R}_3$ are aligned in a "head-to-tail" fashion that is required for the addition of two vectors. Vectors $\mathbf{R}_1$ and $\mathbf{R}_4$ also have that configuration. Finally, note that vectors $\mathbf{R}_3$ and $\mathbf{R}_4$ terminate at the same point of the mechanism (point B). That these two pairs of vectors, when added head to tail, end in the same point indicates that the results of these additions must be equal. Equation (2-1) expresses this observation mathematically.

$$\mathbf{R}_2 + \mathbf{R}_3 = \mathbf{R}_1 + \mathbf{R}_4 \qquad (2\text{-}1)$$

Equation (2-1) indicates that the displacement achieved when vectors $\mathbf{R}_2$ and $\mathbf{R}_3$ are added is the same as that achieved when vectors $\mathbf{R}_1$ and $\mathbf{R}_4$ are added. This vector loop equation must be satisfied by this mechanism, regardless of the pose it takes, as long as it remains assembled as a mechanism.

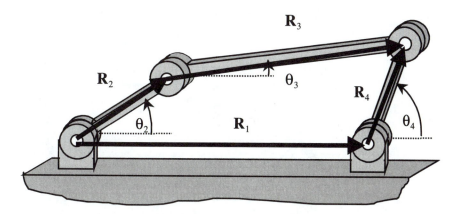

FIGURE 2-3 Displacement vectors drawn along the links of the four-bar mechanism. Note that the link angles are the same as the angles of the displacement vectors relative to the coordinate frame.

2.4 Derivatives of Vectors

The vector loop equation derived in the previous section is a compact, elegant expression of the simple fact the various links of the mechanism are constrained in some manner. In other words, since the four-bar is a single degree-of-freedom mechanism, the arbitrary specification of any one joint angle automatically specifies the other two through the vector equation (2-1). Since a large portion of the analyses will include the computation of velocities and accelerations of the mechanism, it is logical that the derivatives with respect to time be taken of this equation.

It should be clear that the vectors are time-varying since, even though the link lengths remain constant, their orientation (and hence vector directions) change as the mechanism moves. In other mechanisms, both the magnitude and directions may change.

It is easier to take these derivatives if the vector loop equation is broken down into its two scalar components, one dealing with x-components and the other with y-components. Recall the definition of the angle of the vector in Figure 2-1. Figure 2-4 shows the graphical representation of the vector loop equation with the angles of the vector drawn in. Note that the angles are defined as counterclockwise positive from the positive x-direction and drawn to the tail of each vector. Note also that the manner in which the vectors are drawn ensures that the vector angle and the angle of the link that vector represents are the same.

If the above convention is rigorously adhered to, the vector loop equation readily breaks down into its two component equations by using the sines and cosines of the vectors' angles.

$$r_2 \cos\theta_2 + r_3 \cos\theta_3 = r_1 \cos\theta_1 + r_4 \cos\theta_4 \tag{2-2}$$

$$r_2 \sin\theta_2 + r_3 \sin\theta_3 = r_1 \sin\theta_1 + r_4 \sin\theta_4 \tag{2-3}$$

Take a moment to consider these equations. For this particular mechanism, the four r values are link lengths and hence remain constant throughout our analyses. There are four link angles, one of which can be taken to be constant and zero since the coordinate

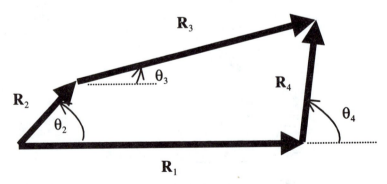

FIGURE 2-4   Vector loop diagram of the four-bar linkage in Figure 2-3. Note that the angles of the vectors (measured from the positive x-axis to the vector at its tail) are the same as the corresponding link angle.

system may be drawn at an arbitrary orientation. This fact allows the problem to be simplified somewhat by drawing the x-axis coincident with one of the vectors. In this example, θ_1 is always zero because it lies along the x-axis.

The other three angles are time-varying, and that must be taken into account when taking the derivatives. Equations (2-4) and (2-5) show the results of taking the first derivative with respect to time of these two component equations. Note that in these equations and throughout the book, we maintain the convention that the time derivative of an angle, θ_i, is represented by the variable ω_i. Likewise, the time derivative of an angular velocity, ω_i, is represented by the angular acceleration α_i.

$$-\omega_2 r_2 \sin\theta_2 - \omega_3 r_3 \sin\theta_3 = -\omega_4 r_4 \sin\theta_4 \tag{2-4}$$

$$\omega_2 r_2 \cos\theta_2 + \omega_3 r_3 \cos\theta_3 = \omega_4 r_4 \cos\theta_4 \tag{2-5}$$

These two equations specify the relationship that must exist among the rotational rates of the three moving links as long as the mechanism remains assembled.

In many analyses, it is assumed that one of the links is caused to move at a constant rotational rate. For example, it can be assumed that link 2 is attached to a large motor that is capable of providing enough torque so that the ω_2 is relatively constant. Under these conditions, ω_2 is said to be the input to the system, and the equations can be rearranged.

$$-\omega_3 r_3 \sin\theta_3 + \omega_4 r_4 \sin\theta_4 = \omega_2 r_2 \sin\theta_2 \tag{2-6}$$

$$\omega_3 r_3 \cos\theta_3 - \omega_4 r_4 \cos\theta_4 = -\omega_2 r_2 \cos\theta_2 \tag{2-7}$$

In traditional kinematic analyses, the velocity solution, which is embodied in equations (2-6) and (2-7), takes place after the position problem is solved. In other words, the values of all the link angles are known at this point in the analysis. Chapter 3 discusses the solution of the position problem in some detail, but at this point it will be assumed that the link angles are known. Therefore, equations (2-6) and (2-7) can be rewritten in matrix form:

$$\begin{bmatrix} -r_3 \sin\theta_3 & r_4 \sin\theta_4 \\ r_3 \cos\theta_3 & -r_4 \cos\theta_4 \end{bmatrix} \begin{bmatrix} \omega_3 \\ \omega_4 \end{bmatrix} = \begin{bmatrix} \omega_2 r_2 \sin\theta_2 \\ -\omega_2 r_2 \cos\theta_2 \end{bmatrix} \tag{2-8}$$

The second derivative of the loop equation is also useful and is easily carried out. One need only keep in mind that the terms in the velocity equations are the products of two time-varying parts (ω and $\cos(\theta)$) and the product rule for differentiation applies.

$$-\alpha_3 r_3 \sin\theta_3 - \omega_3^2 r_3 \cos\theta_3 + \alpha_4 r_4 \sin\theta_4 + \omega_4^2 r_4 \cos\theta_4 = \alpha_2 r_2 \sin\theta_2 + \omega_2^2 r_2 \cos\theta_2 \tag{2-9}$$

$$\alpha_3 r_3 \cos\theta_3 - \omega_3^2 r_3 \sin\theta_3 - \alpha_4 r_4 \cos\theta_4 + \omega_4^2 r_4 \sin\theta_4 = -\alpha_2 r_2 \cos\theta_2 + \omega_2^2 r_2 \sin\theta_2 \tag{2-10}$$

These equations can be put into matrix form, assuming that both the speed and acceleration of the input link are known.

$$\begin{bmatrix} -r_3 \sin\theta_3 & r_4 \sin\theta_4 \\ r_3 \cos\theta_3 & -r_4 \cos\theta_4 \end{bmatrix} \begin{bmatrix} \alpha_3 \\ \alpha_4 \end{bmatrix} = \begin{bmatrix} \alpha_2 r_2 \sin\theta_2 + \omega_2^2 r_2 \cos\theta_2 + \omega_3^2 r_3 \cos\theta_3 - \omega_4^2 r_4 \cos\theta_4 \\ -\alpha_2 r_2 \cos\theta_2 + \omega_2^2 r_2 \sin\theta_2 + \omega_3^2 r_3 \sin\theta_3 - \omega_4^2 r_4 \sin\theta_4 \end{bmatrix}$$

$$\tag{2-11}$$

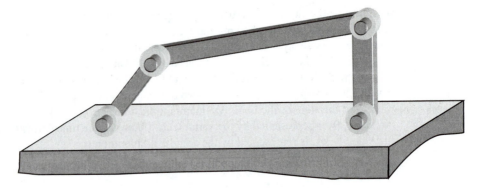

FIGURE 2-5 Four-bar mechanism for Example 2-1.

Example 2-1

Use equation (2-8) to solve for the velocities of links 3 and 4 of the four-bar mechanism shown in Figure 2-5. Use the link lengths and angles shown in Table 2-1. Assume that the velocity of the input link, link 2, is 100 rad/s.

TABLE 2-1 Link Lengths and Angles for Four-Bar Linkage in Example 2-1

Link	Length, r (cm)	Angle, θ (degrees)
1	12.0	0
2	4.0	45
3	10.0	24.65
4	7.0	90.68

The problem is solved by using equation (2-8) in MATLAB. A transcript of the interactive session to solve this problem follows:

```
>> r2=4;
>> r3=10;
>> r4=7;
>> th2=45*pi/180;
>> th3=24.652*pi/180;
>> th4=90.6794*pi/180;
>> %
>> % Note that the angles are in radians
>> %
>> J=[-r3*sin(th3) r4*sin(th4);r3*cos(th3) -r4*cos(th4)]
J =

 -4.1711 6.9995
  9.0886 0.0830
```

```
>> b=[100*r2*sin(th2);-100*r2*cos(th2)]

b =

 282.8427
-282.8427

>> om34=inv(J)*b

om34 =

 -31.319
 21.745
```

So the solution of this position of the mechanism is that link 3 is rotating clockwise at 31.32 rad/s, and link 4 is rotating counterclockwise at 21.74 rad/s.

MATLAB HINT

Trigonometric functions in MATLAB assume that the arguments are in radians. Since most of us are more accustomed to using the unit of degree, you can easily convert between the two by multiplying an angle in degrees by $\pi/180$ to get radians, or conversely, multiply an angle in radians by $180/\pi$ to get degrees. MATLAB makes this a little easier by providing the permanent, global constant, `pi`, which is a numerical representation of π, carried out to the maximum precision of the program.

It is left to the student to verify these results with traditional vector polygon methods.

2.5 Other Common Mechanisms

The proper use of the vector loop equation and its derivatives takes practice. To this end, the following table contains several common mechanisms with consistent vector loop equations along with their first two derivatives. Note that the assignment of a vector loop equation is not unique since the placement of the coordinate frame and vector orientations is somewhat arbitrary. As long as these methods are used in a consistent manner, then the results will be correct. By rederiving these equations, the analytical and mathematical skills required to apply this important technique will be greatly strengthened.

2.6 Vector Chains

Sometimes it is important to explicitly define the motion of a point or points on a mechanism that do not coincide with a joint. Other times, we need to analyze mechanisms that do not form closed loops with the ground link. In these instances, vector

TABLE 2-2 Vector Equations of Four Common Mechanisms

Mechanism	Vector Equation	Velocity Equation	Acceleration Equation
4-Bar	$\mathbf{R}_2 + \mathbf{R}_3 = \mathbf{R}_1 + \mathbf{R}_4$	$\begin{bmatrix} -r_3\sin\theta_3 & r_4\sin\theta_4 \\ r_3\cos\theta_3 & -r_4\cos\theta_4 \end{bmatrix}\begin{bmatrix}\omega_3 \\ \omega_4\end{bmatrix} = \begin{bmatrix}\omega_2 r_2 \sin\theta_2 \\ -\omega_2 r_2\cos\theta_2\end{bmatrix}$	$\begin{bmatrix} -r_3\sin\theta_3 & r_4\sin\theta_4 \\ r_3\cos\theta_3 & -r_4\cos\theta_4 \end{bmatrix}\begin{bmatrix}\alpha_3 \\ \alpha_4\end{bmatrix} = \begin{bmatrix}\alpha_2 r_2 \sin\theta_2 + r_2\omega_2^2\cos\theta_2 + r_3\omega_3^2\cos\theta_3 - r_4\omega_4^2\cos\theta_4 \\ -\alpha_2 r_2\cos\theta_2 + r_2\omega_2^2\sin\theta_2 + r_3\omega_3^2\sin\theta_3 - r_4\omega_4^2\sin\theta_4\end{bmatrix}$
Slider Crank (in-line)	$\mathbf{R}_2 + \mathbf{R}_3 = \mathbf{R}_1$	$\begin{bmatrix}1 & r_3\sin\theta_3 \\ 0 & -r_3\cos\theta_3\end{bmatrix}\begin{bmatrix}\dot{r}_1 \\ \omega_3\end{bmatrix} = \begin{bmatrix}-\omega_2 r_2\sin\theta_2 \\ \omega_2 r_2\cos\theta_2\end{bmatrix}$	$\begin{bmatrix}1 & r_3\sin\theta_3 \\ 0 & -r_3\cos\theta_3\end{bmatrix}\begin{bmatrix}\ddot{r}_1 \\ \alpha_3\end{bmatrix} = \begin{bmatrix}-\alpha_2 r_2\sin\theta_2 - r_2\omega_2^2\cos\theta_2 - r_3\omega_3^2\cos\theta_3 \\ \alpha_2 r_2\cos\theta_2 - r_2\omega_2^2\sin\theta_2 - r_3\omega_3^2\sin\theta_3\end{bmatrix}$
Slider Crank (offset)	$\mathbf{R}_2 + \mathbf{R}_3 = \mathbf{R}_1 + \mathbf{R}_4$	$\begin{bmatrix}1 & r_3\sin\theta_3 \\ 0 & -r_3\cos\theta_3\end{bmatrix}\begin{bmatrix}\dot{r}_1 \\ \omega_3\end{bmatrix} = \begin{bmatrix}-\omega_2 r_2\sin\theta_2 \\ \omega_2 r_2\cos\theta_2\end{bmatrix}$	$\begin{bmatrix}1 & r_3\sin\theta_3 \\ 0 & -r_3\cos\theta_3\end{bmatrix}\begin{bmatrix}\ddot{r}_1 \\ \alpha_3\end{bmatrix} = \begin{bmatrix}-\alpha_2 r_2\sin\theta_2 - r_2\omega_2^2\cos\theta_2 - r_3\omega_3^2\cos\theta_3 \\ \alpha_2 r_2\cos\theta_2 - r_2\omega_2^2\sin\theta_2 - r_3\omega_3^2\sin\theta_3\end{bmatrix}$
Quick Return Mechanism	$\mathbf{R}_1 + \mathbf{R}_3 = \mathbf{R}_2$	$\begin{bmatrix}\cos\theta_3 & -r_3\sin\theta_3 \\ \sin\theta_3 & r_3\cos\theta_3\end{bmatrix}\begin{bmatrix}\dot{r}_3 \\ \omega_3\end{bmatrix} = \begin{bmatrix}-\omega_2 r_2\sin\theta_2 \\ \omega_2 r_2\cos\theta_2\end{bmatrix}$	$\begin{bmatrix}\cos\theta_3 & -r_3\sin\theta_3 \\ \sin\theta_3 & r_3\cos\theta_3\end{bmatrix}\begin{bmatrix}\ddot{r}_3 \\ \alpha_3\end{bmatrix} = \begin{bmatrix}-\alpha_2 r_2\sin\theta_2 - r_2\omega_2^2\cos\theta_2 + 2\dot{r}_3\omega_3\sin\theta_3 + r_3\omega_3^2\cos\theta_3 \\ \alpha_2 r_2\cos\theta_2 - r_2\omega_2^2\sin\theta_2 - 2\dot{r}_3\omega_3\cos\theta_3 + r_3\omega_3^2\sin\theta_3\end{bmatrix}$

loops are not possible, but similar relationships can be derived by specifying the vector chains which relate the important coordinates with the link coordinates. In this section, we explore two important uses of vector chains: the two-link planar robot and specifying the location of a point on the coupler link of a four-bar mechanism.

2.6.1 Two-Link Planar Robot

Figure 2-6 shows a schematic diagram of a two-link planar robot. This example is often used in the robotics literature as the simplest form of a robot made up of two links and two revolute joints. Note that the two joint angles, θ_1 and θ_2, are measured to the base of each link, as previously defined. Note also that the coordinate system is placed at the joint between the ground (link 1) and the first robot link. In this example, it is important that we relate the linkage coordinates, θ_1 and θ_2, to the location of the end effector, given by the vector, $\mathbf{R}_E$. It's easy to see that the vector drawn from the center of the co-ordinate system to the center of the robot gripper is the sum of the two-link displacement vectors.

$$\mathbf{R}_E = \mathbf{R}_2 + \mathbf{R}_3 \tag{2-12}$$

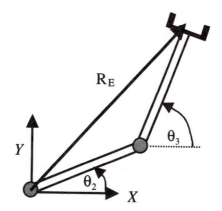

FIGURE 2-6 Two-link manipulator mechanism showing two degrees of freedom and a displacement vector drawn to the end effector.

There exists an important distinction between the vector $\mathbf{R}_E$ shown above and the previously defined displacement vectors. $\mathbf{R}_E$ is not attached to, or moving with, one of the links in the mechanism. In this case, the vector simply indicates the position of the point in the center of the robot gripper. In cases such as these, the angle of the vector is of little significance and we represent the x- and y-components simply as x_E and y_E.

$$x_E = r_2 \cos\theta_2 + r_3 \cos\theta_3 \tag{2-13}$$

$$y_E = r_2 \sin\theta_2 + r_3 \sin\theta_3 \tag{2-14}$$

As before, we can easily take the derivatives of these scalar equations and put the resulting equations in matrix form to find the relationship known as the Jacobian trans-formation.

$$\begin{bmatrix} \dot{x}_E \\ \dot{y}_E \end{bmatrix} = \begin{bmatrix} -r_2 \sin\theta_2 & -r_3 \sin\theta_3 \\ r_2 \cos\theta_2 & r_3 \cos\theta_3 \end{bmatrix} \begin{bmatrix} \omega_2 \\ \omega_3 \end{bmatrix} \tag{2-15}$$

As a final observation, it should be noted that the robotics literature uses a slightly different angle convention, leading to a different form of the Jacobian transformation. In robotics, joint angles are used, which differ from link angles in that they are measured relative to the previous link. This is done because the sensors used in robotic systems measure this relative angle directly. On the other hand, the convention in kinematics literature is to use the absolute angle.

2.6.2 Vector Chains to Describe Motion of an Arbitrary Point

In Figure 2-7, a four-bar linkage is shown in which the coupler link, link 3, carries a point labeled C. There are occasions in which it is important to describe the motion of such a point using vectors. One common application of this method would be to describe "coupler curves." Coupler curves are the curves traced by particular points on the coupler of a four-bar mechanism during its motion. Another common usage of this technique is to describe the constraints between the link angles and the motion of the center of mass (COM) of a particular link. In either case, the approach is the same.

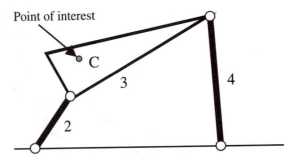

FIGURE 2-7   Four-bar linkage with designated point of interest on link 3.

The first step is to define a displacement vector to the point of interest on the link itself. This vector should originate at the same vertex where the link's displacement vector starts. In our examples to date, the vertex between the crank (link 2) and the coupler (link 3) is the location. Figure 2-8 below shows the correct displacement vector.

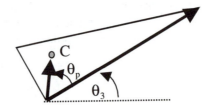

FIGURE 2-8   Displacement vector indicating the position of the point of interest.

Note that two angles are shown in Figure 2-8. Angle θ_3 is defined relative to the horizontal, or x-axis, as was shown previously. The second angle, θ_p, is the angle of the new displacement vector relative to the link's displacement vector. This angle is constant and does not change value while the link is moving. On the other hand, the absolute angle of orientation of the displacement vector is the sum of the link's angle and the relative angle of the newly defined displacement vector.

Figure 2-9 shows the mechanism with the vectors drawn in the appropriate locations.

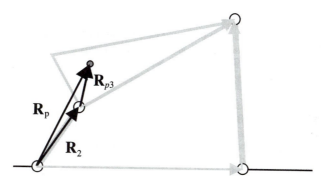

FIGURE 2-9 Four-bar mechanism showing the vectors required to track the motion of a point that is not specifically at one of the joints.

The vector equation that is used to track the location of the point can be easily written.

$$\mathbf{R}_p = \mathbf{R}_2 + \mathbf{R}_{p3} \qquad (2\text{-}16)$$

Keeping in mind the angle that was defined in Figure 2-8, the equation can be broken down into x- and y-components.

$$x_p = r_2 \cos\theta_2 + r_{p3} \cos(\theta_3 + \theta_{p3}) \qquad (2\text{-}17)$$

$$y_p = r_2 \sin\theta_2 + r_{p3} \sin(\theta_3 + \theta_{p3}) \qquad (2\text{-}18)$$

If it is desired that the velocity of the point be known, then simply taking the derivative of these equations leads to the following equations. Note that, while θ_3 is time-varying, θ_{p3} is not, so it does not contribute to the time derivative.

$$\dot{x}_p = -r_2\omega_2 \sin\theta_2 - r_{p3}\omega_3 \sin(\theta_3 + \theta_{p3}) \qquad (2\text{-}19)$$

$$\dot{y}_p = r_2\omega_2 \cos\theta_2 + r_{p3}\omega_3 \cos(\theta_3 + \theta_{p3}) \qquad (2\text{-}20)$$

And similarly, the accelerations can be tracked as well:

$$\ddot{x}_p = -r_2\alpha_2 \sin\theta_2 - r_2\omega_2^2 \cos\theta_2 - r_{p3}\alpha_3 \sin(\theta_3 + \theta_{p3}) - r_{p3}\omega_3^2 \cos(\theta_3 + \theta_{p3}) \qquad (2\text{-}21)$$

$$\ddot{y}_p = r_2\alpha_2 \cos\alpha_2 - r_2\omega_2^2 \sin\theta_2 + r_{p3}\alpha_3 \cos(\theta_3 + \theta_{p3}) - r_{p3}\omega_3^2 \sin(\theta_3 + \theta_{p3}) \qquad (2\text{-}22)$$

2.7 Summary

This chapter presented a review of vector loop equations and their derivatives for the analysis of mechanical mechanisms. The vector loop equation is a compact and elegant representation of the constraints imposed on the mechanism's movement by the joints between the links. Displacement vectors are associated with each link in the mechanism and oriented in such a way that the magnitude of the vector is the link length and the angle of the vector corresponds to the angle of the link. This procedure is modified slightly for sliding contact joints, as will be shown in Chapter 4, with a slider-crank mechanism.

By breaking the vector loop equation down into x- and y-components, the derivatives of the equation can easily be found, thus providing information relating velocities and accelerations of the links. These derivative equations will form the foundation for both kinematic and dynamic simulations of the mechanisms.

CHAPTER 2 PROBLEMS

1. Formulate the velocity and acceleration equations for a three-link planar robot.

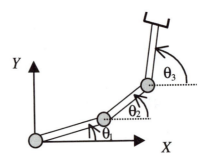

2. Formulate the velocity and acceleration equations for a three-link planar robot using relative angles as shown below.

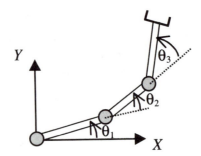

3. Consider the Geneva mechanism shown below. Write the vector loop equation appropriate for this mechanism, and derive the velocity and acceleration equations.

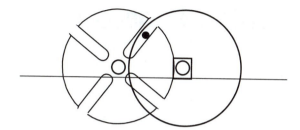

4. Consider the Scotch Yoke shown below. Write the vector loop equation appropriate for this mechanism, and derive the velocity and acceleration equations.

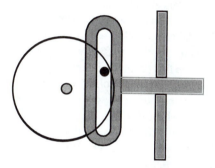

5. In the double-slider shown in the following figure, assign an appropriate coordinate system, write a vector loop equation, and derive the relationship between the motion of the vertical slider, 2, with the angular motion of the connecting link and the velocity of the horizontal slider.

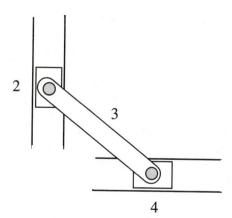

CHAPTER 3

SOLUTIONS OF THE POSITION PROBLEM

3.1 Overview

In the course of mechanism analysis, the first step is the so-called position solution. For a single degree-of-freedom mechanism the problem answers the question: If I know the position of one of the links (relative to the ground) then what are the positions of the other links in the mechanism? As was shown in Chapter 2, the vector loop equation can be used to solve this problem. For example, equations (2-2) and (2-3) contain sufficient information to allow one to solve for θ_3 and θ_4, given θ_2 and all of the link lengths. Unfortunately, the equations are nonlinear, transcendental equations in θ_3 and θ_4 and are therefore not easily solved. In this chapter, a numerical method for the solution of the position problem will be presented and the implementation of this method in MATLAB program scripts will be discussed. It should be noted that this chapter is included for the sake of completeness only. Many kinematics textbooks discuss this method, and MATLAB is particularly well-suited for the solution of the equations derived here. On the other hand, the material discussed in this chapter is not used in the subsequent chapters and can be safely skipped without loss of continuity.

3.2 Numerical Solutions of Nonlinear Algebraic Equations

The Newton-Raphson method for the solution of nonlinear equations is briefly presented here, although a more thorough discussion can be found elsewhere.[1] Simply stated, the Newton-Raphson method is an iterative method for the solution of nonlinear algebraic equations that starts with an initial guess at the unknowns and incrementally improves the guess until the solution is "close enough" to the correct solution. The incremental improvements are computed by using a series expansion of the nonlinear equations and the concept of "close enough" is usually defined in the context of numerical tolerance and engineering applications.

Without loss of generality, the problem is shown for two equations with two unknowns, though it is easily expanded as needed. Assume that the problem is expressed in the following form:

---

[1]Mabie, H. H., and C. F. Reinholtz, *Mechanisms and Mechanics of Machinery,* 4th ed., New York: John Wiley & Sons, 1987.

$$f_1(q_1, q_2) = 0 \tag{3-1}$$

$$f_2(q_1, q_2) = 0 \tag{3-2}$$

where q_1 and q_2 are the unknowns to be found. The first step is to express the variables in terms of an estimation of the solution (denoted by an overbar) and a small correction factor (denoted by a Δ prefix) that is the difference between the estimation and the solution.

$$q_i = \bar{q}_i + \Delta q_i \tag{3-3}$$

The Taylor Series expansion is a well-known mathematical tool to represent non-linear function of variables that have been decomposed as shown in equation (3-3). The expansion says that a function can be represented as an infinite series starting with the function evaluated at the estimate ($\bar{q}_1, \bar{q}_2$). The expansion for the first function is shown below:

$$f_1(q_1, q_2) = f_1(\bar{q}_1, \bar{q}_2) + \frac{\partial f_1}{\partial q_1}\bigg|_{\bar{q}_1, \bar{q}_2} \Delta q_1 + \frac{\partial f_1}{\partial q_2}\bigg|_{\bar{q}_1, \bar{q}_2} \Delta q_2 + [\text{higher-order terms}] \tag{3-4}$$

The higher-order terms are usually neglected, because the series will be used in an iterative fashion to approximate the function and it is desirable that the equation be linear in the delta-terms, which would not be the case if higher-order terms were included. It's important to note that the partial derivative terms in equation (3-4) are evaluated at the estimated values of the unknowns and are therefore computable to a simple numerical value. They become numerical coefficients in the expression. Applying the expansion to both functions, we can represent the results in matrix form:

$$\begin{bmatrix} f_1(\bar{q}_1, \bar{q}_2) \\ f_2(\bar{q}_1, \bar{q}_2) \end{bmatrix} + \begin{bmatrix} \dfrac{\partial f_1}{\partial q_1}\bigg|_{\bar{q}_1, \bar{q}_2} & \dfrac{\partial f_1}{\partial q_2}\bigg|_{\bar{q}_1, \bar{q}_2} \\ \dfrac{\partial f_2}{\partial q_1}\bigg|_{\bar{q}_1, \bar{q}_2} & \dfrac{\partial f_2}{\partial q_2}\bigg|_{\bar{q}_1, \bar{q}_2} \end{bmatrix} \begin{bmatrix} \Delta q_1 \\ \Delta q_2 \end{bmatrix} = \begin{bmatrix} 0 \\ 0 \end{bmatrix} \tag{3-5}$$

Examination of equation (3-5) leads to a method of computing the difference between an estimate of the unknowns ($\bar{q}$'s) and the correct angle. Solving equation (3-5) for the difference (Δq's):

$$\begin{bmatrix} \Delta q_1 \\ \Delta q_2 \end{bmatrix} = \begin{bmatrix} \dfrac{\partial f_1}{\partial q_1}\bigg|_{\bar{q}_1, \bar{q}_2} & \dfrac{\partial f_1}{\partial q_2}\bigg|_{\bar{q}_1, \bar{q}_2} \\ \dfrac{\partial f_2}{\partial q_1}\bigg|_{\bar{q}_1, \bar{q}_2} & \dfrac{\partial f_2}{\partial q_2}\bigg|_{\bar{q}_1, \bar{q}_2} \end{bmatrix}^{-1} \begin{bmatrix} -f_1(\bar{q}_1, \bar{q}_2) \\ -f_2(\bar{q}_1, \bar{q}_2) \end{bmatrix} \tag{3-6}$$

Solution of this class of nonlinear problem is then reduced to the method shown in the flow chart of Figure 3-1.

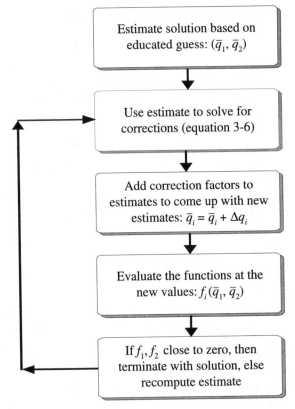

FIGURE 3-1 Schematic representation of Newton-Raphson algorithm applied to the solution of the position problem in kinematics.

3.3 The Position Problem of a Four-Bar Linkage

In the previous chapter, the vector loop equation for a four-bar linkage was derived and it was broken down into x- and y-components—see equations (2-2) and (2-3). Taking advantage of the fact that a reference frame can always be chosen to align with the base link (link 1) and rearranging the equations, the following two functions are defined:

$$f_1(\theta_3, \theta_4) = r_2 \cos\theta_2 + r_3 \cos\theta_3 - r_1 - r_4 \cos\theta_4 = 0 \tag{3-7}$$

$$f_2(\theta_3, \theta_4) = r_2 \sin\theta_2 + r_3 \sin\theta_3 - r_4 \sin\theta_4 = 0 \tag{3-8}$$

Now the position problem can be stated as follows: For a given set of link lengths and for a given value of θ_2, find those values of θ_3 and θ_4 for which functions f_1 and f_2 evaluate to zero. Since f_1 and f_2 are nonlinear and transcendental (the unknowns are contained within transcendental functions), linear matrix methods do not apply. The Newton-Raphson method for solving nonlinear equations works well for this problem.

First, redefine the variables in terms of some nominal value of the answer, which is considered "close" to the answer and the correction factor that makes up the difference:

$$\theta_3 = \overline{\theta}_3 + \Delta\theta_3 \tag{3-9}$$

$$\theta_4 = \overline{\theta}_4 + \Delta\theta_4 \tag{3-10}$$

where θ_3 and θ_4 represent the solution to the problem, the overbars indicate the nominal values that are close to the solution. and the Δ's indicate the correction factors. Using the Taylor Series as described above, and putting the results in the form of equation (3-6), the following matrix equation is obtained:

$$\begin{bmatrix} \Delta\theta_3 \\ \Delta\theta_4 \end{bmatrix} = \begin{bmatrix} -r_3\sin\overline{\theta}_3 & r_4\sin\overline{\theta}_4 \\ r_3\cos\overline{\theta}_3 & -r_4\cos\overline{\theta}_4 \end{bmatrix}^{-1} \begin{bmatrix} -f_1(\overline{\theta}_3,\overline{\theta}_4) \\ -f_2(\overline{\theta}_3,\overline{\theta}_4) \end{bmatrix} \tag{3-11}$$

3.4 MATLAB Solution of the Position Problem of a Four-Bar Linkage

The MATLAB computational environment is well-suited to the solution of the position problem as outlined above. The following function shows one possible implementation of the use of MATLAB for solving the nonlinear transcendental equations that arise in a typical position problem.

MATLAB HINT

MATLAB allows you to keep groups of commands that may be repeatedly executed in separate text files. The filenames for these have the extension ".m" and hence are known as "m-files." (Imagine if the programmers had chosen the extension ".x"!) There are two distinct types of m-files. The script file is simply a collection of commands that can be executed from the command prompt by typing the name of the file. These commands execute exactly as they would if typed individually. The second type of m-file is the function file that operates like a function in C or a subroutine in FORTRAN. The function accepts arguments and returns values, and while executing, the function does not have access to the variables in the work-space (unless they have been defined as global). When writing a function file, the first line must have the format shown below, with the "function" keyword, the re-turn variables, and the arguments of the function.

---

posso14.m

```
function [th3, th4]=posso14(th,rs)
%posso14 Position solution of a 4-bar mechanism
%
% Script used to implement Newton-Raphson
% method for solving nonlinear
% position problem of a 4-bar mechanism.
%
%Copyright 2001
%John F. Gardner
%
%
```

```
%
%inputs of function
%
%th(1) = theta-2
%th(2) = theta-3-bar (starting guess)
%th(3) = theta-4-bar (starting guess)
%
%rs(1) = r-1
%rs(2) = r-2
%rs(3) = r-3
%rs(4) = r-4
%
th2=th(1);
th3bar=th(2);
th4bar=th(3);
%
% set condition for convergence
%
epsilon=1.0E-6;
%Intialize the f-vector
%
% Compute the functions as a two-element vector:
%
f=[rs(3)*cos(th3bar)-rs(4)*cos(th4bar)+rs(2)*cos(th2)-
 rs(1);rs(3)*sin(th3bar)-rs(4)*sin(th4bar)+rs(2)*sin(th2)];
%
% Repeatedly compute the correction factors
% as per equation (3-11)
%
while norm(f)>epsilon
 J=[-rs(3)*sin(th3bar) rs(4)*sin(th4bar);
 rs(3)*cos(th3bar) -rs(4)*cos(th4bar)];
 dth=inv(J)*(-1.0*f);
 th3bar=th3bar+dth(1);
 th4bar=th4bar+dth(2);
 f=[rs(3)*cos(th3bar)-rs(4)*cos(th4bar)+rs(2)*cos(th2)-
 rs(1);rs(3)*sin(th3bar)-rs(4)*sin(th4bar)+rs(2)*sin(th2)];
 norm(f)
end;
th3=th3bar;
th4=th4bar;
```

MATLAB HINT

MATLAB is based on computer code and numerical methods written in the 1970s specifically for matrix manipulation and inversion. Hence, the fundamental data structure is a matrix, and matrix operations are easily performed. This function demonstrates this fact by solving the matrix formula with a single line with simple syntax.

As an example, consider the four-bar first encountered in Example 2-1. The following is a record of a MATLAB session in which the function is used to solve for the position:

```
» rs(1)=12;
» rs(2)=4;
» rs(3)=10;
» rs(4)=7;
» th(1)=0;
» th(2)=45*pi/180;
» th(3)=135*pi/180;
» possol4(th,rs)

ans =
0.7688    1.6871
```

which corresponds to θ_3 of 44.05 degrees and θ_4 of 96.66 degrees.

MATLAB HINT

Floating-point numbers in MATLAB are internally represented with a large number of significant digits. By default, MATLAB will show only five significant digits in the command window. For some applications (such as setting initial conditions for simulations), more precise representations may be required. You can adjust the number of significant digits displayed using the FORMAT command in the MATLAB window. Typing FORMAT LONG at the >> prompt will allow you to see the entire representation.

3.5 Position Solutions and Initial Guesses

In the previous section, a MATLAB function was introduced that—given link lengths, input angle, and initial guesses—will provide the angles of the other two links in a four-bar mechanism. In this section, some subtleties of such an approach are discussed.

First, it should be realized that many mechanisms have multiple solutions for the position problem. For example, Figure 3-2 shows two different ways to assemble a four-bar linkage of given link lengths and input angle. The solution that will be provided by the Newton-Raphson scheme is, in general, dependent on the initial guess. An appropriate choice of initial guess will ensure proper operation of the algorithm.

The best way to come up with initial guesses is to sketch the linkage approximately to scale. An "eyeball" estimate of the unknown angles is almost always good enough. Also, it is often required that the position solution be obtained for a number of positions, usually equal intervals of crank rotation around a complete circle. This can easily be done sequentially using a MATLAB script file. In this case, only the first initial guess need be provided. For all other positions, the last solved position can be used because it

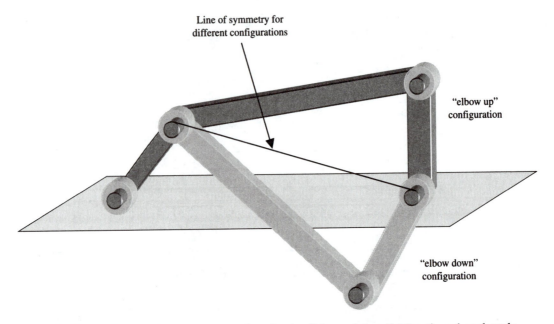

FIGURE 3-2 Two possible configurations for a four-bar linkage of given link lengths and crank angle.

will be necessarily close to the new position. The following example demonstrates this procedure.

Example 3-1

Using the link lengths for the four-bar mechanism described in Example 2-1, solve the position problem (find angles θ_3 and θ_4) for a full rotation of the input link (link 2), at 5° intervals. Plot the angles θ_3 and θ_4 versus θ_2.

First, a sketch of the mechanism, as shown in Figure 3-3, allows the user to estimate the unknown angles for an input angle of 0°. From this figure, values of 45° for θ_3 and 100° for θ_4 are estimated. The following MATLAB script sets up a loop in which the

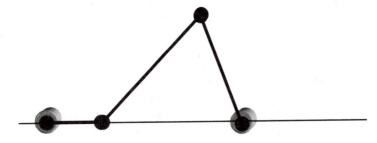

FIGURE 3-3 Sketch of the initial configuration for the four-bar mechanism in Example 3-1. Note that link 2 is shown at an angle of zero, while link 3 appears to be at approximately 45° and link 4 is approximately 100°.

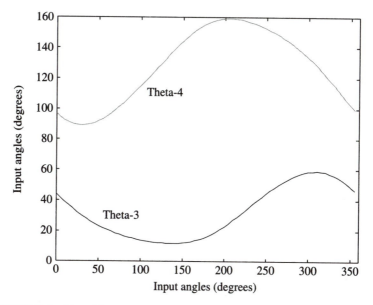

FIGURE 3-4 Plot of θ_3 and θ_4 as a function of input angle for the four-bar linkage under consideration in Example 3-1.

input angle, θ_2, is incremented from 0 to 360 in increments of 5. At each value of the angle, the function possol4( ) (described previously) is called to solve for the unknown angles. After the script is executed, MATLAB plotting commands are used to generate the plot shown in Figure 3-4.

MATLAB HINT

The following example shows the use of both script and function m-files. The file shown below is a script file which calls the function file shown earlier in this chapter. Using a function m-file to encapsulate a particular function and using a script to repeatedly call that function is a powerful combination.

---

fourbar.m

```
%fourbar Sequential position solution of a four-bar linkage
%
%Script file to solve the position solution for 5-degree
%increments
%of input link angle, for a full rotation of the mechanism.
%
%Copyright 2001
%John F. Gardner
%
%
```

```
% Handy conversion constant
%
D2R = pi/180.0;
%
% Set up link angles
%
rs(1) = 12.0;
rs(2) = 4.0;
rs(3) = 10.0;
rs(4) = 7.0;
%
% Set up the input and initial guesses
%
th(1) = 0.0;
th(2) = 45 * D2R;
th(3) = 100 * D2R;
%
% define the increment
%
dth = 5 * D2R;
%
for i = 1:72
  ths = possol4(th,rs);
  %Store results in a matrix, in degrees
  angles(i,:) = [th(1)/D2R ths(1)/D2R ths(2)/D2R];
  th(1) = th(1) + dth;
  th(2) = ths(1);
  th(3) = ths(2);
end
```

Note that the script stored the results in a matrix called `angles` (3 columns and 72 rows) that can now be plotted.

```
» plot(angles(:,1),angles(:,2),angles(:,1),angles(:,3))
» axis([0 360 0 160])
» xlabel('Input angles (degrees)')
» ylabel('Solved angles (degrees)')
» text(110,110,'Theta-4')
» text(50,35,'Theta-3')
```

MATLAB HINT

One of the most useful and versatile commands in the MATLAB environment is the `plot` command. The online help facility describes `plot` in detail, but we see it here in its simplest form. The `plot` command requires two vectors: the first representing the x-coordinates of the data set, the second containing the y-coordinates. Often, data describing many physical variables can be stored in a single matrix, with each column dedicated to each variable. In this case, we make use of the `:` operator to select a given column for the `plot` command. For example, the command `plot(mydata(:,1),mydata(:,2))` will plot the first two columns in the matrix `mydata` against one another. It is sometimes useful to read the colon operator in a situation such as this as meaning "all rows." In this manner, the `plot` command could be recited as: plot all rows of the first column of `mydata` against all rows of the second column of `mydata`.

3.6 Summary

In this chapter, the vector loop equation was used to set up the position problem for mechanisms. Using the Newton-Raphson approach, a MATLAB function and script file was introduced to solve the position problem for a four-bar linkage. These routines are easily modified for different mechanism configurations.

CHAPTER 3 PROBLEMS

1. Examine θ_3 and θ_4 in the plot shown in Figure 3-4. Can you conclude anything about the velocities of links 3 and 4 from these plots? Carefully explain your answer.

2. Using the function file that accompanies this text, perform the analysis in Example 3-1 for several different lengths r_2. In particular, gradually increase the value until the function no longer returns a value (the numerical algorithm will fail to converge). Explain why this happens.

3. Modify the m-file `possol4.m` to keep track of the number of iterations required for each solution. Use the MATLAB plot command to plot the number of iterations for each solution versus crank angle. What can you conclude from this result?

4. Change the convergence tolerance, epsilon, to `1.0E-9` and repeat problem 3.

CHAPTER 4

KINEMATIC SIMULATIONS USING SIMULINK

4.1 What Is a Kinematic Simulation?

In engineering analysis, the term *computer simulation* has evolved to mean the numerical solution of ordinary differential equations. The field of system dynamics covers the development and solutions of differential equations that describe the time-varying responses of a wide variety of physical systems, and there are many fine textbooks that cover this area in great detail.[1] In this book, a subclass of differential equations will be considered: those that are the result of kinematic analyses.

The term *kinematic simulation* refers to the use of a simulation package to repeatedly solve the kinematic constraint equations of a mechanism and to integrate the resultant velocities (or accelerations) so that the positions (and velocities) can be computed through numerical integration. In this manner, a simulation package, such as Simulink, offers several advantages over the traditional method of solving for the motion of a mechanism. The most obvious advantage is that the position problem is solved implicitly by the simulation package. The user need only solve the position problem for one pose of the mechanism to provide appropriate initial conditions for the simulation.

A kinematic simulation can be used to compute and plot the velocities of the links of a mechanism or can be extended, as will be shown, to compute accelerations as well. In both cases, the starting point is the vector loop equation reviewed in Chapter 2. In this chapter, kinematic simulations will be discussed in which both constant-speed and accelerating input links are modeled.

4.2 Velocity Solution via Kinematic Simulation

4.2.1 Vector Loop Equations for the Slider Crank

To demonstrate this method, a slider-crank mechanism will be simulated through several complete rotations of the crank at constant speed. Figure 4-1 shows a schematic of a slider-crank mechanism as it might be found in a single cylinder four-stroke engine.

---

[1]Shearer, L. J., Kulakowski, B. T., and Gardner, J. F., *Dynamic Modeling and Control of Engineering Systems,* 2nd ed., Englewood Cliffs, NJ: Prentice Hall, 1997.

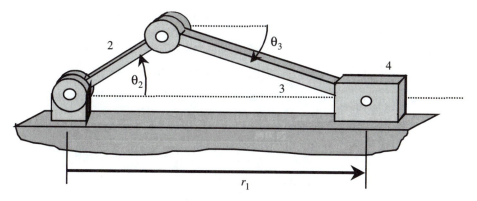

FIGURE 4-1 Schematic of a slider-crank mechanism showing the numbers of each link and the angles of each.

Although this mechanism has only one degree of freedom (DOF), the motion of all three moving links will be considered in the simulation. The simulation will take ω_2 as input and compute ω_3 and $\dot{r}_1$ (by the same method described in Example 2-1).

Figure 4-2 shows the vector loop that describes the slider crank. Note an important difference between this vector loop and the one describing the four-bar (Figure 2-3). In the previous case, all vectors had constant magnitude but varying orientation. In this case, vector $\mathbf{R}_1$ has a time-varying magnitude and constant orientation. This is an important point when taking the time derivatives.

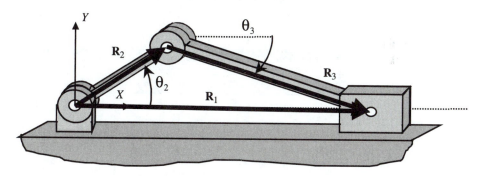

FIGURE 4-2 Vector loop for the slider crank.

The vector loop equation for the slider crank is

$$\mathbf{R}_2 + \mathbf{R}_3 = \mathbf{R}_1 \tag{4-1}$$

which can be broken down into x- and y-coordinates:

$$r_2 \cos\theta_2 + r_3 \cos\theta_3 = r_1 \tag{4-2}$$

$$r_2 \sin\theta_2 + r_3 \sin\theta_3 = 0 \tag{4-3}$$

Next, take the derivatives of these equations with respect to time.

$$-r_2\omega_2 \sin\theta_2 - r_3\omega_3 \sin\theta_3 = \dot{r}_1 \tag{4-4}$$

$$r_2\omega_2 \cos\theta_2 + r_3\omega_3 \cos\theta_3 = 0 \tag{4-5}$$

where $\dot{r}_1$ is the rate of change of the magnitude of the $\mathbf{R}_1$ vector. This is also the translational speed of the slider relative to the ground.

Equations (4-4) and (4-5) can be rearranged in matrix form as shown:

$$\begin{bmatrix} r_3 \sin\theta_3 & 1 \\ -r_3 \cos\theta_3 & 0 \end{bmatrix} \begin{bmatrix} \omega_3 \\ \dot{r}_1 \end{bmatrix} = \begin{bmatrix} -r_2\omega_2 \sin\theta_2 \\ r_2\omega_2 \cos\theta_2 \end{bmatrix} \tag{4-6}$$

Equation (4-6) is a statement of the velocity problem for the slider crank if the crank speed, ω_2, is known.

4.2.2 Simulink Simulation of the Slider-Crank Kinematics

Equation (4-6) can be used to solve for ω_3 and $\dot{r}_1$ when ω_2 and all positions are known. If ω_2 is considered an input to the simulation, then numerical integration can be used to compute θ_2, θ_3, and r_1 from the velocities. In the block-oriented graphical approach of Simulink, this will require three integration blocks to start the construction of the simulation. Figure 4-3 shows a Simulink window with these integrator blocks.

To proceed, the inputs to the integrators must be appropriately connected. The first input is ω_2, which is assumed to be constant and is the input to the simulation. Any one of a number of blocks from the SOURCES library can be used; in this case, a constant block will be chosen. The other two velocities can be computed from the vector

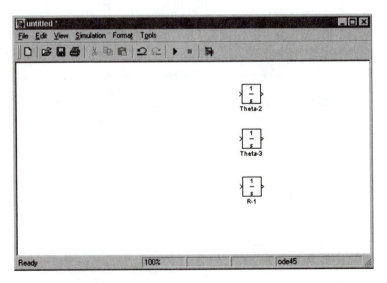

FIGURE 4-3 Simulink window with the beginnings of the kinematic simulation. Integrator blocks are labeled with the name of the signal, which is the *result* of the simulation.

loop equation (4-6). A MATLAB function is written to solve this equation. The function is called `compvel` and is stored in the file `compvel.m`. The content of the file is shown below.

compvel.m

```
function [x]=compvel(u)
%
%Copyright 2001
%John F. Gardner
%
%
% function to compute the unknown velocities for
% a slider crank with constant crank input
%
%  u(1) = omega-2
%  u(2) = theta-2
%  u(3) = theta-3
%
%  Define the geometry
%
r2=1.0;
r3=4.0;
%
a=[r3*sin(u(3)) 1 ;-r3*cos(u(3)) 0];
b=[-r2*u(1)*sin(u(2));r2*u(1)*cos(u(2))];
%
x=inv(a)*b;
```

MATLAB HINT

MATLAB allows you to keep groups of commands that may be repeatedly executed in separate text files. The filenames for these have the extension ".m" and hence are known as "m-files." (Imagine if the programmers had chosen the extension ".x"!) There are two distinct types of m-files. The script file is simply a collection of commands that can be executed from the command prompt by typing the name of the file. These commands execute exactly as they would if typed individually. The second type of m-file is the function file which operates like a function in C or a subroutine in FORTRAN. The function accepts arguments and returns values and while executing, the function does not have access to the variables in the workspace (unless they have been defined as global). When writing a function file, the first line must have the format shown below, with the "function" keyword, the return variables, and the arguments of the function.

Note that the link lengths of the slider crank, r_2 and r_3, are defined within this function. This function can be embedded into a Simulink simulation by using the MATLAB `function` block, found in the "Functions and Tables" library. The `function` block can take vectors as inputs and outputs, but the vectors must be "assembled" from the

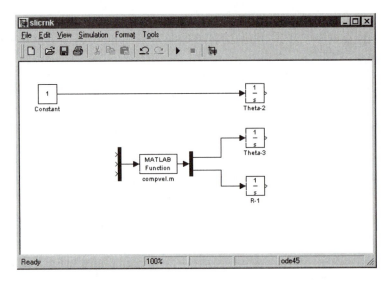

FIGURE 4-4   Expanding the block diagram to include the function `compvel.m` to connect the inputs of the integrators to outputs of the function.

original signals. Figure 4-4 shows that the input vector (**u** in the function) is assembled using a `Mux` (multiplexor) block. Note that the order in which the signals are connected is the same as the order used in the function. Likewise, the output vector (**x** in the function) is disassembled using a `DeMux` block.

Next, the appropriate signals are "hooked up" to the `Mux` block. The results are shown in Figure 4-5.

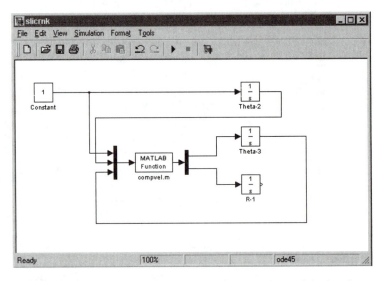

FIGURE 4-5   Connect the appropriate signals to the input of the function.

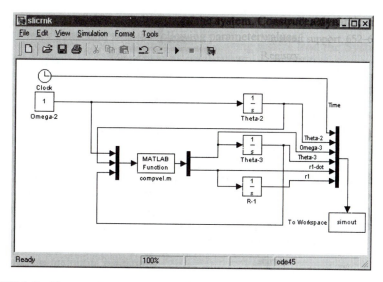

FIGURE 4-6 The complete kinematic simulation for a constant input speed. The angles and velocities are stored in the MATLAB workspace in the variable `yout` (file: `slicrnk.mdl`).

To complete the simulation, a `Mux` block is added to collect the outputs to a matrix that will be available in the MATLAB workspace when the simulation is complete. Note that you can also label the signal lines in Simulink. Double-click on the line, and a text box will open that will allow you to enter a signal name. This is particularly valuable for interpreting and documenting the simulation. The completed simulation is shown in Figure 4-6. The `clock` block shown in Figure 4-6 is not needed if version 5.x or higher of MATLAB is used; it is shown here for the sake of completeness.

4.2.3 Establishing Initial Conditions

The preceding section showed how one can construct a block-diagram-oriented simulation of the slider crank based on vector-loop equations. Before the simulation can be run, however, the appropriate initial conditions for the integrators must be established. This is a critical step in the solution of any differential equations; however, in the case of kinematic simulations, it is particularly important. If a consistent set of initial conditions is not used, the simulation will fail. In the case of the simulation shown in Figure 4-6, the initial values for θ_2, θ_3, and r_1 must be such that they represent an actual position of the mechanism.

In this case, the initial conditions are solved with simple geometry. To make the job easier, it will be assumed that the initial position of the slider crank will be $\theta_2 = 0$, which puts the crank in line with the connecting rod. This position is known as "top dead center" (TDC) in piston-cylinder arrangements. By inspection, it can be seen that θ_3 is also equal to zero and the slider is at a position that is $r_2 + r_3$ units from the origin of the coordinate system. The consistent set of initial conditions is shown in Table 4-1.

TABLE 4-1   Initial Conditions for Simulation

Position	Value
θ_2	0 radians
θ_3	0 radians
r_1	5.0 inches

These can be made available to the simulation in one of two methods. The more straightforward method is to double-click on the integrator icons in the block diagram and enter the numbers directly on the "Initial condition" line. Unfortunately, this method can be quite awkward for more complicated initial conditions, as shall be shown in subsequent sections. The other method is to refer to variables that are established in the MATLAB workspace. This two-step process is easily demonstrated below.

At the MATLAB prompt, type:

```
>> th20 = 0;
>> th30 = 0;
>> r10 = 5.0;
```

Double-clicking on the integrator icon will bring up the dialog box shown in Figure 4-7. Note that the variable name is entered in the "Initial condition:" line in the dialog box. For the time being, the other elements of the dialog box can be ignored.

The reasons for this indirect method in establishing the initial conditions are twofold. First, it allows the user to alter the initial conditions easily by a series of commands within the MATLAB workspace, instead of having to double-click on each and

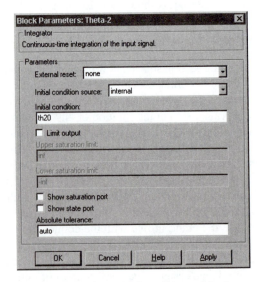

FIGURE 4-7   Dialog box to set up intitial conditions of the integrator in Simulink.

every integrator every time the initial conditions are to be changed. Second, it allows Simulink to have access to the initial conditions with the maximum range of precision allowed within MATLAB. For the case shown here, this is not important because the initial conditions are known precisely. However, as will be seen shortly, most simulations require the use of more complex geometries, and the precision of the initial conditions can be important for proper performance of the simulation.

4.2.4 Simulation Results

Now the simulation will be used to analyze the motion of the mechanism. It will be assumed that the crank operates at 1800 rpm (188.5 rad/s). (This is a common speed for single-cylinder lawn mower engines.) At this speed, it completes two revolutions in just under 0.07 seconds, so the simulation will run from 0 to 0.07 seconds. To set the crank speed, double-click on the constant block in the simulation. The resulting dialog box allows you to enter an arbitrary value for the crank speed. It is important to note that the units used for angular velocity is rad/s, and crank speeds in rev/min (rpm) must be converted to rad/s for the purposes of these analyses.

In MATLAB version 5.x, the default simulation parameters (algorithm, step size, tolerance) work fine here. In earlier versions of MATLAB, you'll want to make the minimum step size something on the order of $1E-5$ and the tolerance about the same. Note that the simulation has no graph or Scope blocks, so that when the simulation is run, nothing will appear on the screen. When the end of the simulation is indicated, however, the MATLAB workspace contains a new matrix, simout. One can then use the MATLAB plot command to look at the system motion. The matrix simout has six columns (one for every variable entering the Mux block, and as many rows as there were time steps in the simulation. Referring back to Figure 4-6, we see that the six columns, in order, are: Time, θ_2, ω_3, θ_3, $\dot{r}_1$, and r_1. Note that in MATLAB 5.x, the time variable is automatically stored in the workspace variable tout. To see a plot of the piston displacement, r_1, versus time, the following MATLAB command can be used:

```
>> plot(simout(:,1),simout(:,6))
>> xlabel('Time (sec)')
>> ylabel('Piston Displacment (in)')
```

MATLAB HINT

One of the most useful and versatile commands in the MATLAB environment is the plot command. The online help facility describes plot in detail, but we see it here in its simplest form. The plot command requires two vectors, the first representing the x-coordinates of the data set, the second containing the y-coordinates. Often, data describing many physical variables can be stored in a single matrix, with each column dedicated to each variable. In this case, we make use of the : operator to select a given column for the plot command. For example, the command plot(mydata(:,1),mydata(:,2)) will plot the first two columns in the matrix mydata against one another. It is sometimes useful to read the colon

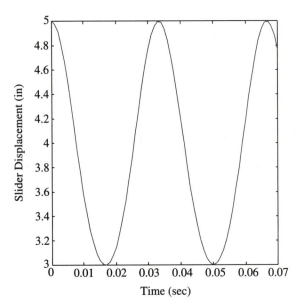

FIGURE 4-8 Plot of piston displacement versus time for simulation of slider crank.

operator in a situation such as this as meaning "all rows." In this manner, the `plot` command could be recited as: plot all rows of the first column of `mydata` against all rows of the second column of `mydata`.

Similarly, the rotational speed of the connecting rod can be viewed by issuing the following commands:

```
>> plot(simout(:,1),simout(:,3))
>> xlabel('Time (sec)')
>> ylabel(' Connecting Rod Speed (rad/s)')
```

The plot is shown in Figure 4-9.

Finally, it might be interesting to look at the piston velocity. This information would be particularly useful to a designer of an air compressor or piston engine who is concerned with the friction characteristics between the piston rings and the cylinder wall.

```
>> plot(simout(:,1),simout(:,5))
>> xlabel('Time (sec)')
>> ylabel('Piston Speed (in/s)')
```

The results of this are shown in Figure 4-10.

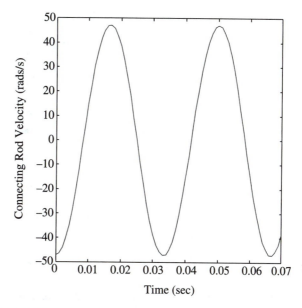

FIGURE 4-9 Plot of connecting rod angular velocity versus time
for simulation of slider-crank mechanism.

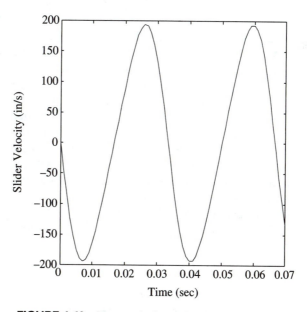

FIGURE 4-10 Piston velocity versus time for simulation
of slider-crank mechanism.

4.3 Acceleration Solution via Kinematic Simulation

4.3.1 Including Acceleration in the Simulation

Keeping the example of the slider crank, recall the velocity equations that were derived by taking the first time derivative of the vector loop equations (4-4) and (4-5). When taking the second time derivative, care must be exercised. In this case there are products of time-varying quantities, and they must be handled appropriately. The second derivative of the vector loop equations for the slider crank are shown below:

$$-r_2\dot{\omega}_2 \sin\theta_2 - r_2\omega_2^2 \cos\theta_2 - r_3\dot{\omega}_3 \sin\theta_3 - r_3\omega_3^2 \cos\theta_3 = \ddot{r}_1 \tag{4-7}$$

$$r_2\dot{\omega}_2 \cos\theta_2 - r_2\omega_2^2 \sin\theta_2 + r_3\dot{\omega}_3 \cos\theta_3 - r_3\omega_3^2 \sin\theta_3 = 0 \tag{4-8}$$

In the context of a simulation, the displacements (θ_2, θ_3, and r_1) are considered knowns (since they are the results of integration). Likewise, when the simulation is extended to include accelerations, the velocities (ω_2, ω_3, and $\dot{r}_1$) will also be considered knowns. In the simulation being assembled, the angular acceleration of the input link will be considered the input to the simulation. Finally, the convention of using α_i to represent the angular acceleration will be adopted. With this in mind, the acceleration equations can be expressed in compact matrix form as shown.

$$\begin{bmatrix} r_3 \sin\theta_3 & 1 \\ -r_3 \cos\theta_3 & 0 \end{bmatrix} \begin{bmatrix} \alpha_3 \\ \ddot{r}_1 \end{bmatrix} = \begin{bmatrix} -(r_2\alpha_2 \sin\theta_2 + r_2\omega_2^2 \cos\theta_2 + r_3\omega_3^2 \cos\theta_3) \\ r_2\alpha_2 \cos\theta_2 - r_2\omega_2^2 \sin\theta_2 - r_3\omega_3^2 \sin\theta \end{bmatrix} \tag{4-9}$$

Comparing equation (4-9) with the matrix form of the velocity equation (4-6) points out an important feature of the derivatives of the vector loop equations. The 2×2 matrices on the left side of both equations are identical. This can be used as an important checkpoint when deriving the equations for new mechanisms.

Much of the structure of the new simulation is the same as the simulation that was assembled previously. Starting with the latest version of the simulation (Figure 4-6) the signal lines for the velocities are removed, as is the input block. The blocks are also rearranged slightly to make room for the additional integrators needed to integrate the accelerations to velocities. Figure 4-11 shows the first step in the modification.

A new function file, `compacc.m`, will be required to solve the acceleration equation (4-9) above. This new function will require additional inputs of α_2 (the simulation input) ω_2, and ω_3. The `Mux` block is then extended, and three new `Integration` blocks are added to the simulation. Figure 4-12 shows the modified simulation.

The output vector has also been changed. The variable `simout` now contains six columns: time, θ_2, ω_3, θ_3, $\dot{r}_1$, and r_1.

The new function, `compacc.m`, has a form very similar to the `compvel.m` file from Chapter 4 and is shown below.

---

comacc.m

```
function [x]=compacc(u)
%
% function to compute the unknown accelerations for
% a slider crank with variable speed crank input
%
```

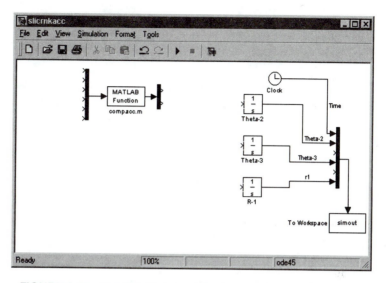

FIGURE 4-11 Simulink diagram of kinematic simulation of slider crank showing the first step in extending the simulation to acceleration analysis.

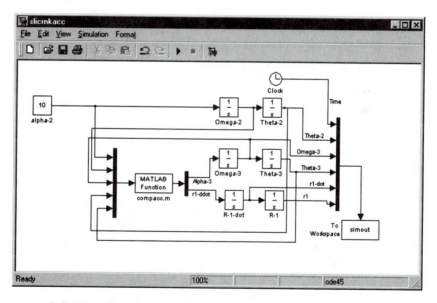

FIGURE 4-12 Simulink simulation for slider crank with accelerations and variable speed crank (file: slicrnkacc.mdl).

```
%  u(1) = alpha-2
%  u(2) = omega-2
%  u(3) = omega-3
%  u(4) = theta-2
%  u(5) = theta-3
%
%  Define the geometry
%
r2=1.0;
r3=4.0; (cont....)
%
a=[r3*sin(u(5)) 1 ;-r3*cos(u(5)) 0];
b=[-(r2*u(1)*sin(u(4))+r2*u(2)^2*cos(u(4))+r3*u(3)^2*cos(u(5)));
 r2*u(1)*cos(u(4))-r2*u(2)^2*sin(u(4))-r3*u(3)^2*sin(u(5)) ];
%
x=inv(a)*b;
```

4.3.2 Running the Slider-Crank Simulation

Previously, it was noted that care must be taken to ensure that the consistent initial conditions are established. For the velocity-based simulation, that meant that the initial displacements must represent a valid position of the linkage. In addition to that requirement, the initial velocities in this simulation must also represent a valid combination of velocities. A good way to think of this is to realize that the slider crank is a single degree-of-freedom mechanism. If the speed of the crank is specified (initial velocity for ω_2), then the other two velocities (ω_3 and $\dot{r}_1$) are also specified through the velocity equations (equation 4-6).

Two examples are given here. First, the simulation will be executed from a zero-velocity state under steady acceleration. Second, the simulation will be executed for a constant speed condition.

Example 4-1

The initial conditions for this case are similar to that of the velocity-based simulation with the exception that the velocity ICs are all set to zero. The input (α_2) is specified with a Simulink constant block. The initial conditions are specified in Table 4-2.

TABLE 4-2 Initial Conditions for Constant Acceleration Simulation

Integrator	Initial Condition
θ_2	0 radians
θ_3	0 radians
r_1	5.0 inches
ω_2	0 rad/s
ω_3	0 rad/s
$\dot{r}_1$	0 in.s

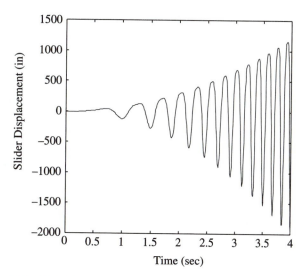

FIGURE 4-13 Linear acceleration experienced by the piston during the constant acceleration simulation.

The input acceleration is set to 10 rad/s/s, and the simulation is run for 4 seconds of simulated time. At the end of 4 seconds, the crank speed should be 40 rad/s (about 380 rpm), and the crank will have rotated a total of 80 radians (nearly 13 revolutions). Figure 4-13 shows a plot of the acceleration experienced by the piston $\ddot{r}_1$ predicted by the simulation. Note that the number of cycles shown here match with the predicted number of 13 revolutions. Note also that the acceleration is decidedly not sinusoidal in nature. This clearly demonstrates that the slider in the slider crank does not exhibit simple harmonic motion.

Example 4-2

The initial conditions for the constant-speed simulation is a bit more challenging because one must first solve the velocity problem for the initial pose of the mechanism to provide the necessary initial conditions for the simulation. As an example, the method will be demonstrated for the same speed as was simulated earlier, 188.5 rad/s. Obviously, the initial condition for the ω_2 integrator is specified (188.5). To find the other two initial conditions, the velocity problem must be solved for this one position (see Example 2-1).

```
>> th20=0;
>> th30=0;
>> r10=5.0;
>> om20=188.5;
>> r2=1.0;
>> r3=4.0;
```

```
>> a=[r3*sin(th30) 1; -r3*cos(th30) 0 ]

a =

  0 1
  -4 0

>> b=[-r2*om20*sin(th20); r2*om20*cos(th20)]

b =

  0
  188.5000

>> x=inv(a)*b

x =

  -47.1250
  0

>> om30=x(1);
>> r1dot0=x(2);
```

The appropriate initial conditions are summarized in Table 4-3.

TABLE 4-3 Initial Conditions for Constant Speed Simulation

Integrator	Workspace Variable	Initial Condition
θ_2	th20	0
θ_3	th30	0 rad
r_1	r10	5.0 in
ω_2	om20	188.5 rad/s
ω_3	om30	−47.125 rad/s
$r1$-dot	r1dot0	0.0 in/s

As before, the simulation was run for 0.07 seconds. Figure 4-14 shows the slider acceleration for this run.

It's interesting to take a moment to interpret this particular result. The mechanism is running with a crank speed of 1800 rpm (typical for a small gasoline engine), the crank length is 1 inch, and the connecting rod length is 4 inches. Nonetheless, the peak acceleration experienced by the piston is over 40,000 in/s/s. This corresponds to an acceleration of over 100 g's. Therefore, the dynamic force experienced by the piston is over 100 times the weight of the piston. Clearly, dynamic forces are significant for mechanisms of even modest size and speed. In the following chapters, dynamic simulations will be introduced, which will allow the direct solutions of the forces within the mechanism.

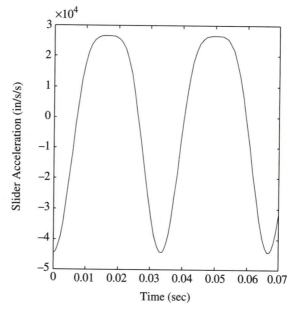

FIGURE 4-14 Acceleration experienced by slider for Example 4-2.

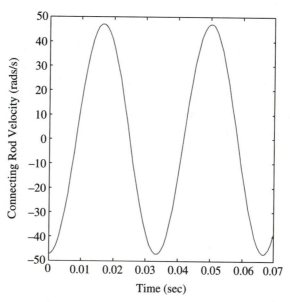

FIGURE 4-15 Plot of connecting rod velocity versus time for constant speed run.

Finally, consider the velocity of the connecting rod, ω_3 (Figure 4-15). Since this simulation considered constant-speed crank motion, the result should be the same as was found for the velocity-only kinematic simulation, shown in Figure 4-9. The two plots are identical, as expected.

4.4 The Consistency Check

One of the most interesting features of a kinematic simulation is that it computes redundant data. The most rudimentary analysis of a slider crank, for instance, demonstrates that it is a single degree-of-freedom mechanism, requiring only one coordinate to completely determine the position of all its links at an instant in time. On the other hand, the kinematic simulation independently computes θ_2, θ_3, and r_1. This apparent contradiction is resolved when we realize that the vector loop equations form the basis of solving the velocities or accelerations of the mechanism. Therefore, the integrals of those variables must also satisfy the vector loops. By checking the degree to which the resulting positions satisfy the original vector loop equation, we get an indication of the validity of the simulation itself.

A new function is introduced that takes the displacement variables as input and returns a value that represents the error inherent in the computation. For the slider-crank mechanism, recall that the vector loop equation is:

$$\mathbf{R}_1 - \mathbf{R}_2 - \mathbf{R}_3 = 0 \qquad (4\text{-}10)$$

We then can define the error in a vector sense as:

$$\mathbf{E} = \mathbf{R}_1 - \mathbf{R}_2 - \mathbf{R}_3 \qquad (4\text{-}11)$$

Then the norm of the $\mathbf{E}$ vector is a scalar that indicates the validity of the simulation. Typically, if there is an error in the coding of the problem—either in the structure of the simulation diagram or in the function—then the error will start out large and quickly grow without bound, an indication that further debugging is required. On the other hand, if the error begins at zero and grows more slowly but still becomes excessive within the timeframe required for the simulation (typically about 10 rotations of the mechanism), then the numerical integration is not converging adequately. To solve this problem, lower the error limits in Simulink. Finally, if the error starts at a non-zero value and grows steadily, then your initial conditions were not self-consistent and should be recomputed.

The new function, `comperr`, is shown below.

---

comperr.m

```
function e=comperr(u)
%
%
% function to compute the displacement error
% inherent in the kinematic simulation
%
% u(1) - theta-2
% u(2) - theta-3
% u(3) - r1
%
r2 = 1.0;
r3 = 4.0;
%
```

```
ex = u(3) - r2*cos(u(1)) - r3*cos(u(2));
ey = -r2*sin(u(1)) - r3*sin(u(2));
%
e=norm([ex ey]);
```

The function is easily incorporated into either a velocity-based or an acceleration-based simulation, as Figure 4-16 shows.

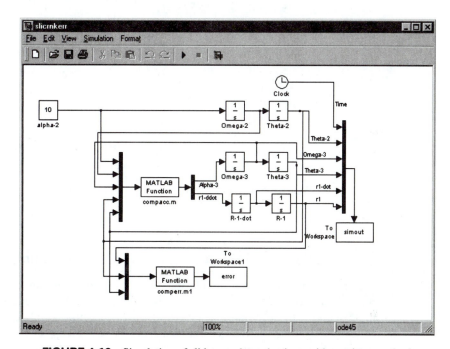

FIGURE 4-16   Simulation of slider-crank mechanisms with consistency check
(file: `slicrnkerr.mdl`).

After this modification has been made, the simulation is run again. Now there are two matrices in the workspace, `simout` and `error`. To see how the error changes throughout the run:

```
>> plot(simout(:,1),error)
>> xlabel('Time (sec)')
>> ylabel('RMS Consistency Error (in)')
```

The result is shown in Figure 4-17. Note that the consistency error is nearly constant and quite low.

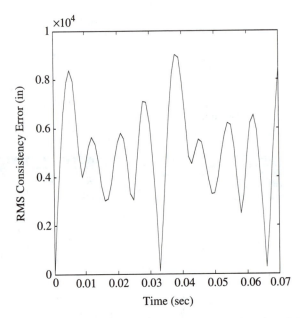

FIGURE 4-17   Plot of root-mean-square error as a function of time
for simulation of slider-crank mechanism.

4.5   Kinematic Simulation of a Four-Bar Mechanism

To further illustrate the concept of kinematic simulations, the slider-crank simulation
will be modified to simulate a four-bar mechanism. This is easily done with only minor
modifications to the Simulink model. The two function files, `compacc.m` and
`conchk.m`, must be modified to reflect the proper vector loop equations. In Chapter 2,
the vector loop equations for the four-bar were presented. Equations (2-6) and (2-7)
show the first derivatives of the *x*- and *y*-components of the equation. Taking the second
derivative yields:

$$
\begin{bmatrix} -r_3\sin\theta_3 & r_4\sin\theta_4 \\ r_3\cos\theta_3 & -r_4\cos\theta_4 \end{bmatrix}\begin{bmatrix} \alpha_3 \\ \alpha_4 \end{bmatrix} = \begin{bmatrix} r_2\alpha_2\sin\theta_2 + r_2\omega_2^2\cos\theta_2 + r_3\omega_3^2\cos\theta_3 - r_4\omega_4^2\cos\theta_4 \\ -r_2\alpha_2\cos\theta_2 + r_2\omega_2^2\sin\theta_2 + r_3\omega_3^2\sin\theta_3 - r_4\omega_4^2\sin\theta_4 \end{bmatrix}
$$

$$(4\text{-}12)$$

The new function, entitled `compacc4.m`, is listed below.

comacc4.m

```
function [x]=compacc4(u)
%compacc4 computes accelerations of a four-bar mechanism
%
% function used in kinematic simulation of 4-bar
% mechanism.
%
%Copyright 2001
%John F. Gardner
```

```
%
%  u(1) = alpha-2
%  u(2) = omega-2
%  u(3) = omega-3
%  u(4) = omega-4
%u(5) = theta-2
%  u(6) = theta-3
%  u(7) = theta-4
%
%  Define the geometry
%
r1 = 12.0; r2 = 4.0;
r3 = 10.0; r4 = 7.0;
%
a = [-r3*sin(u(6)) r4*sin(u(7)) ;r3*cos(u(6)) -r4*cos(u(7))];
%
b(1) = r2*u(1)*sin(u(5))+r2*u(2)^2*cos(u(5))+r3*u(3)^2*cos(u(6))-
r4*u(4)^2*cos(u(7));
%
b(2) = -2*u(1)*cos(u(5))+r2*u(2)^2*sin(u(5))+r3*u(3)^2*sin(u(6))-
r4*u(4)^2*sin(u(7));
%
x=inv(a)*b';
```

The Simulink diagram requires only slight modification, as seen in Figure 4-18.

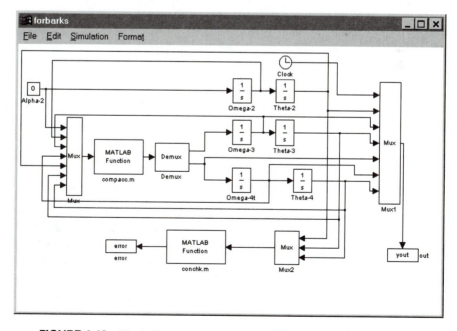

FIGURE 4-18 Block diagram of kinematic simulation for a four-bar mechanism
(file: `fourbarks.mdl`).

To run this simulation, a consistent set of initial conditions must be provided, as discussed previously. Fortunately, the position solution for a four-bar has already been presented in Chapter 3, and the velocity solution is easily performed using the vector loop equations. The following example illustrates the use of the kinematic simulation of the four-bar mechanism.

Example 4-3

Examine the accelerations and velocities of links 3 and 4 for a four-bar linkage having the following dimensions:

Parameter	Value
r_1	120 cm
r_2	40 cm
r_3	100 cm
r_4	70 cm
$\theta_2 \, (t = 0)$	0 rad
ω_2 (assumed constant)	−250 rad/s

First, the MATLAB routine `possol4`, which was introduced in Chapter 3, will be used to find the other two angles (θ_3 and θ_4) that correspond to this initial condition.

```
>> rs = [120 40 100 70]
rs =
 120 40 100 70
>> th=[0 45*pi/180 135*pi/180]

th =

   0   0.7854   2.3562

>> possol4(th,rs)

ans =

   0.7688   1.6871
```

The default in MATLAB is to show only five significant digits of the answer, although it is computed to far greater precision. Because of the cumulative nature of round-off error, these simulations are among the few engineering applications in which it makes sense to carry as many significant digits as possible. To get full precision from MATLAB, issue the following commands:

```
>> format long
>> ans
```

```
ans =

0.76879354899128    1.68712997846810
```

Then use cut and paste to insert these answers into the initial conditions box in the simulation for the θ_3 and θ_4 integrators.

Finally, solve the velocity problem for this position ($\theta_2 = 0$) of the linkage and ω_2 = −250 rad/s as shown in the table. Equation (2-8) shows the vector-matrix form of the velocity solution. Following the method demonstrated in Example 2-1, MATLAB is again used to find ω_3 and ω_4 as consistent initial conditions. At the end of the previous set of commands, θ_3 and θ_4 were saved in the generic ans vector in the workspace.

```
>> th3=ans(1);
>> th4=ans(2)

>> a=[-rs(3)*sin(th3)   rs(4)*sin(th4)
rs(3)*cos(th3)   -rs(4)*cos(th4)]

a =

 -69.52686081652185    69.52686081652185
 71.87499999999999     8.12500000000001

>> b=[-250*rs(2)*sin(0);250*rs(2)*cos(0)]

b =

         0
     10000

>> inv(a)*b

ans =

    125
    125
```

These initial conditions are easily implemented.

At a crank speed of 250 rad/s, the mechanism makes a complete revolution in 0.025 seconds. If the simulation is allowed to run for 0.1 seconds, then approximately 4 revolutions will be simulated. Figure 4-19 is a plot of angular acceleration versus time for links 3 and 4 of the mechanism. The four complete revolutions are clearly visible in this plot.

Similarly, Figure 4-20 shows the angular velocities of the two links for this simulation run.

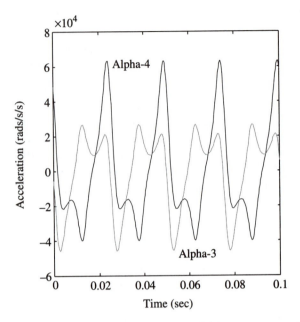

FIGURE 4-19 Angular accelerations of links 3 and 4 for constant speed operation of the four-bar linkage in Example 4-3.

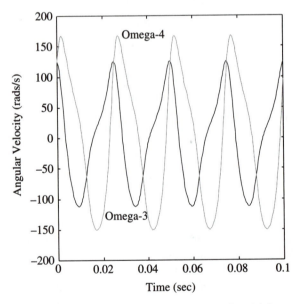

FIGURE 4-20 Angular velocities of links 3 and 4 for the four-bar mechanism in Example 4-3.

4.6 Summary

The concept of a kinematic simulation was introduced in this chapter. A kinematic simulation is a simulation in which the inputs are either speed or acceleration of one of the input links. Equations derived from the vector loop equation are used to solve for the remaining accelerations or velocities of the linkage. Those accelerations and velocities are then integrated with respect to time to produce velocities and positions. An important aspect of kinematic simulations is the appropriate assignment of initial conditions to the integrators. Typically, this requires that the engineer solve the position and perhaps the velocity problem by hand for at least one position of the mechanisms. As will be seen in the next chapter, it is essential that these initial conditions be known to a high degree of precision because small errors in round-off are quickly magnified by the numerical integration routines.

CHAPTER 4 PROBLEMS

1. For a four-bar mechanism with the following link lengths:

    $r_1 = 10.0$ cm
    $r_2 = 3.0$ cm
    $r_3 = 8.0$ cm
    $r_4 = 6.0$ cm

    Implement a kinematic simulation of the mechanism, and run it for several revolutions of the crank (link 2), using a constant crank speed of 6000 rpm. Use the simulation to answer the following questions.
    a. What are the initial conditions for this simulation?
    b. What is the maximum angular acceleration experienced by link 4?
    c. What is the maximum velocity experienced by the bearing between links 2 and 3?
    d. What is the maximum velocity experienced by the bearing between links 3 and 4?
    e. What is the maximum angular acceleration experienced by link 4?
    f. What is the maximum translational velocity experienced by a point located at the midpoint of link 4? In what direction is the acceleration oriented?

2. For the four-bar mechanism described in Problem 1, simulate the mechanism for the case in which it begins with a crank speed of 6000 rpm and experiences a constant angular velocity of link 2 equal to -15 rad/s^2.

3. Consider an inverted slider crank as shown in the figure at the top of page 52. Implement a kinematic simulation for the $r_1 = 0.5$ m, $r_4 = 0.72$ m. Plot the translational acceleration experienced by the slider for one revolution of the crank, assuming a constant crank speed of 200 rad/s. Generate a family of plots for five different values of r_4 ranging from 0.5 m to 1.0 m. What conclusions do you draw from these plots?

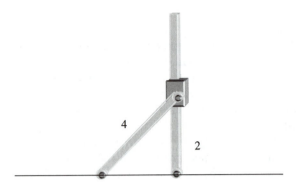

4. A "coupler curve" is the path described by a single point moving with link 3 (the coupler) of a four-bar mechanism. Modify the kinematic simulation used in Problem 1 to trace the coupler curve of an arbitrary point on the coupler. Use MATLAB to draw the coupler curve for the mechanism described in Problem 1, with a coupler point located at the midpoint of link 3. Draw a family of coupler curves for coupler points located on the perpendicular, which passes through the midpoint of link 3 (see figure below).

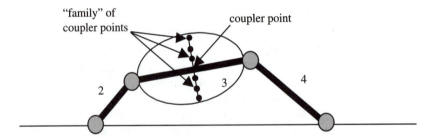

5. An *offset* slider crank is a slider-crank mechanism in which the axis of motion of the slider does not pass through the center of rotation of the crank. Derive the vector loop equation, and take the appropriate derivatives so that you can implement a kinematic simulation that includes acceleration. How does the simulation of an offset slider crank differ from one that is not offset?

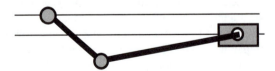

CHAPTER 5

INTRODUCING DYNAMICS

5.1 Overview

In this chapter, we will lay the foundation for the most powerful computer-based technique introduced in this book: dynamic simulations of constrained systems. Using a relatively simple example, we will build a full dynamic simulation of a slider on an inclined plane with a pendulum attached (see Figure 5-1) and examine its behavior under a variety of conditions. With this simple example, we introduce the simultaneous constraint method for dynamic simulations in which kinematic constraints and Newton-Euler equations are solved in a system of simultaneous linear equations to find forces of constraint and accelerations necessary for the simulation.

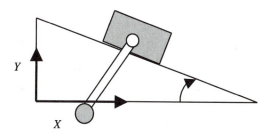

FIGURE 5-1   Slider on an inclined plane with pendulum attached. Note the choice of coordinate system at the corner of the ramp.

5.2 Step 1: Simulating the Slider on Inclined Plane

We proceed with the analysis piece by piece, starting with the slider on the ramp and neglecting the pendulum. Figure 5-2 shows a free-body diagram of the slider. Note that friction (F_f), normal force (N), and weight ($M_s g$) are all shown.

The free-body diagram quickly leads to the following two equations of motion for the slider:

$$N \sin \gamma - F_f \cos \gamma = M_s \ddot{x}_s \tag{5-1}$$

$$N \cos \gamma - M_s g + F_f \sin \gamma = M_s \ddot{y}_s \tag{5-2}$$

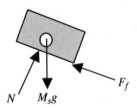

FIGURE 5-2 Free-body diagram of block without pendulum attached.

where (x_s, y_s) denotes the position of the center of mass of the slider in the reference frame.

The friction force, F_f, opposes motion and is proportional to the normal force as shown:

$$F_f = \sigma\mu N \qquad (5\text{-}3)$$

where

$$\sigma = \begin{cases} +1 & \text{for motion down the ramp} \\ -1 & \text{for motion up the ramp} \end{cases}$$

Finally, we note that the two accelerations are not independent. The slope of the ramp provides a constraint for the motion of the slider such that the velocity must always be parallel to the ramp surface. Mathematically, this is expressed in the following equation:

$$\dot{x}_s \tan\gamma = -\dot{y}_s \qquad (5\text{-}4)$$

from which we can easily derive the constraints between the accelerations.

$$\ddot{x}_s \tan\gamma = -\ddot{y}_s \qquad (5\text{-}5)$$

These equations (5-1, 5-2, and 5-5) can be combined to form a system of three equations and three unknowns.

$$\begin{bmatrix} M_s & 0 & \sigma\mu\cos\gamma - \sin\gamma \\ 0 & M_s & -(\sigma\mu\sin\gamma + \cos\gamma) \\ \tan\gamma & 1 & 0 \end{bmatrix} \begin{bmatrix} \ddot{x}_s \\ \ddot{y}_s \\ N \end{bmatrix} = \begin{bmatrix} 0 \\ -M_s g \\ 0 \end{bmatrix} \qquad (5\text{-}6)$$

Figure 5-3 shows the Simulink model of this simple system. The function that computes the accelerations follows.

---

rampslider.m

```
function out=rampslider(u)
%
%  function to compute accelerations of the slider
%  on an inclined plane with friction
%
%  Author:  J.F. Gardner
%
%  Copyright 2001
```

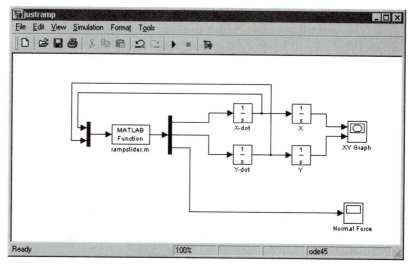

FIGURE 5-3 Simulink model of the slider on the ramp—the first step in modeling the slider-pendulum system in Figure 5-1 (file: `justramp.mdl`).

```
%
% u(1) = x-dot
% u(2) = y-dot
%
m     = 1.0; % kg
gamma = 30 *pi/180; % radians
mu    = 0.5;
g     = 9.8687;
%
% compute the down-ramp velocity to find direction
% of friction force
%%
sdot=u(1)*cos(gamma)-u(2)*sin(gamma);
%
if sdot> 0
   sigma = 1.0;
else
   sigma = -1.0;
end
%
a=[m           0     sigma*mu*cos(gamma)-sin(gamma)  ;
   0           m    -(sigma*mu*sin(gamma)+cos(gamma))  ;
   tan(gamma) 1         0];
%
b=[0;-m*g;0];
%
out=inv(a)*b;
```

As you can see from Figure 5-3, this is a simulation with no input. The initial conditions will determine its response. Running this simulation is not very interesting and merely verifies that the block will slide down the ramp at an acceleration equal to that component of gravity that lies along the incline. This result is easily produced through elementary physics.

On the other hand, the task of adequately modeling the more complicated pendulum-slide configuration of Figure 5-1 is much more difficult.

5.3   Step 2: Adding the Pendulum

Figure 5-4 shows the free-body diagram for the slider block with the forces between the pendulum and the slider shown. Note the convention used for forces that occur between members. F_{psx} is the x-component of the force exerted by the pendulum on the slider block. This convention (noting the source of the force, then the object of the force as ordered subscripts) will be used throughout the text and introduced more formally in Chapter 6.

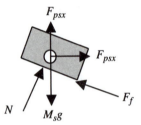

FIGURE 5-4   Free-body diagram of the slider block
with reaction forces from the pendulum shown.

Figure 5-5 shows the free-body diagram of the pendulum itself. Note that the reaction forces from the slider to the pendulum are noted with their subscripts reversed.

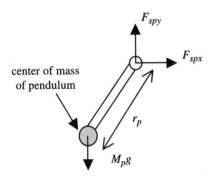

FIGURE 5-5   Free-body diagram of the pendulum.

At this point, we will use the fact that, for two rigid bodies, A and B, joined at a point where the force acting from A on B is equal and opposite to the force acting from B on A. Therefore, we can easily see that $F_{spy} = -F_{psy}$ and $F_{spx} = -F_{psx}$. The equations of motion that follow use these relationships implicitly.

The equations of motion for the slider are changed slightly to account for the pendulum force.

$$N \sin \gamma - \sigma \mu N \cos \gamma - F_{spx} = M_s \ddot{x}_s \tag{5-7}$$

$$N \cos \gamma - M_s g + \sigma \mu N \sin \gamma - F_{spy} = M_s \ddot{y}_s \tag{5-8}$$

Similarly for the pendulum, we derive the following three equations of motion.

$$F_{spx} = M_p \ddot{x}_p \tag{5-9}$$

$$F_{spy} - M_p g = M_p \ddot{y}_p \tag{5-10}$$

$$-r_p \cos \theta_p F_{spy} + r_p \sin \theta_p F_{spx} = I_p \ddot{\theta}_p \tag{5-11}$$

The last equation represents the summation of torques about the center of mass, and θ_p is the angle of the pendulum measured relative to the positive x-axis at the end of the pendulum attached to the slider. This is consistent with the displacement vectors shown in Figure 5-6.

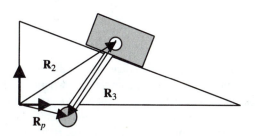

FIGURE 5-6 Displacement vectors assigned to the moving bodies in Figure 5-1.

This leads immediately to the following vector loop equation:

$$\mathbf{R}_2 + \mathbf{R}_3 = \mathbf{R}_p \tag{5-12}$$

The x- and y-components are easily derived. Note that the vectors $\mathbf{R}_2$ and $\mathbf{R}_p$ are shown simply as x and y variables, not with length and angle variables. This is because these two vectors do not correspond to an actual link.

$$x_s + r_p \cos \theta_p = x_p \tag{5-13}$$

$$y_s + r_p \sin \theta_p = y_p \tag{5-14}$$

Taking the derivative twice and regrouping yields the constraint equations for the accelerations of the slider and the pendulum center of mass.

$$\ddot{x}_s - \ddot{x}_p - r_p (\sin \theta_p) \ddot{\theta}_p = r_p \omega_p^2 \cos \theta_p \tag{5-15}$$

$$\ddot{y}_s - \ddot{y}_p + r_p (\cos \theta_p) \ddot{\theta}_p = r_p \omega_p^2 \sin \theta_p \tag{5-16}$$

Finally, for completeness, we repeat the constraint equation that reflects the fact that the slider is free to move only along the surface of the ramp.

$$\ddot{x}_s \tan \gamma = -\ddot{y}_s \tag{5-17}$$

5.4 Step 3: Assembling the Matrix Equation

Equations (5-7) through (5-11) and (5-15) through (5-17) create a system of eight equations that are linear in the following unknowns: $\ddot{x}_s$, $\ddot{y}_s$, $\ddot{x}_p$, $\ddot{y}_p$, $\ddot{\theta}_p$, N, F_{spx}, and F_{spy}. Equation (5-18) below shows the set of equations in matrix form.

$$
\begin{bmatrix}
M_s & 0 & 0 & 0 & 0 & \sigma\mu C_g - S_g & 1 & 0 \\
0 & M_s & 0 & 0 & 0 & -(\sigma\mu S_g + C_g) & 0 & 1 \\
0 & 0 & M_p & 0 & 0 & 0 & -1 & 0 \\
0 & 0 & 0 & M_p & 0 & 0 & 0 & -1 \\
0 & 0 & 0 & 0 & I_p & 0 & -r_p S_p & r_p C_p \\
1 & 0 & -1 & 0 & -r_p S_p & 0 & 0 & 0 \\
0 & 1 & 0 & -1 & r_p C_p & 0 & 0 & 0 \\
S_g & C_g & 0 & 0 & 0 & 0 & 0 & 0
\end{bmatrix}
\begin{bmatrix}
\ddot{x}_s \\
\ddot{y}_s \\
\ddot{x}_p \\
\ddot{y}_p \\
\ddot{\theta}_p \\
N \\
F_{spx} \\
F_{spy}
\end{bmatrix}
=
\begin{bmatrix}
0 \\
-M_s g \\
0 \\
-M_p g \\
0 \\
r_p \omega_p^2 C_p \\
r_p \omega_p^2 S_p 0 \\
\end{bmatrix}
$$

$$(5\text{-}18)$$

Note that this matrix equation contains some of the information that was embedded in the matrix equations encountered in Chapter 4. In addition, the equations of motion augment the kinematic constraints. If we embed this matrix equation in a simulation in which the accelerations are integrated, then the velocities and displacements would be available to compute the matrix and the right side of the equation.

In this manner, we can compute the accelerations of this system as well as the forces of constraint simultaneously. There are more compact descriptions of the dynamics of this particular system, but this one can be derived with sophomore-level vector mechanics, and the forces of constraint—which can be very important for detailed dynamic analysis and design purposes—are directly computed.

In the next section, we create a Simulink simulation to predict the dynamics of the block and pendulum system.

5.5 Step 4: Creating a Dynamic Simulation

Figure 5-7 shows the Simulink simulation that embodies the dynamics and kinematics derived in the previous two sections. Note that the five accelerations that were unknowns in the matrix equation are now integrated twice, and the results are plotted to the screen through various display blocks.

The following m-file function is used in the simulation to set up and solve the matrix equation. Note that the parameters of the system are defined within the function.

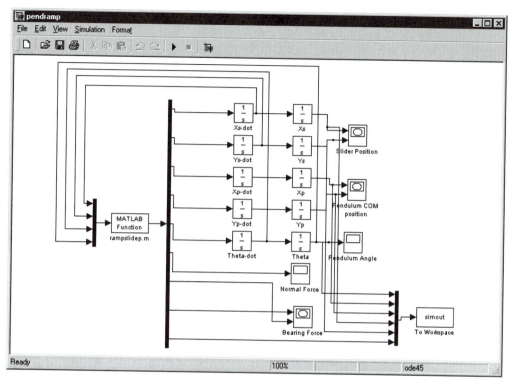

FIGURE 5-7 Simulink model of the block/pendulum system on an inclined plane (file: `pendramp.mdl`).

rampsliderp.m

```
function out=rampslidep(u)
%
%  function to compute accelerations of the slider
%  and pendulum on an inclined plane with friction
%
% Author:  J.F. Gardner
%
% Copyright 2001
%
% u(1) = x-dot
% u(2) = y-dot
% u(3) = omega-p
% u(4) = theta-p
%
ms    = 5.0; % kg
gamma = 30 *pi/180; % radians
mu    = 0.5;
g     = 9.8687;
rp    = 0.5;  % m
```

```
mp  = 0.5;  % kg
Ip  = 0.01; % kg m^2
%
% compute the down-ramp velocity to find direction
% of friction force
%
sdot=u(1)*cos(gamma)-u(2)*sin(gamma);
%
if sdot> 0
   sigma = 1.0;
else
   sigma = -1.0;
end
%
Cp = cos(u(4));
Sp = sin(u(4));
a=zeros(8,8);
a(1,1) = ms;  a(1,6) = sigma*mu*cos(gamma)-sin(gamma);  a(1,7) = 1;
a(2,2) = ms;  a(2,6) = -(sigma*mu*sin(gamma)+cos(gamma))  ;a(2,8) = 1;
a(3,3) = mp;  a(3,7)=-1;
a(4,4) = mp;  a(4,8)=-1;
a(5,5) = Ip;  a(5,7)=-rp*Sp;  a(5,8)=rp*Cp;
a(6,1) = 1;  a(6,3) = -1;  a(6,5) = -rp*Sp;
a(7,2) = 1;  a(7,4)=-1;   a(7,5)=rp*Cp;
a(8,1) = sin(gamma);  a(8,2)=cos(gamma);
%
b=[0;-ms*g;0;-mp*g;0;rp*u(3)^2*Cp;rp*u(3)^2*Sp;0];
%
out=inv(a)*b;
out=[out; sdot];
```

5.6 Step 5: Setting Initial Conditions and Running Simulation

Up to this point, all of the analysis has been general, applying to any circumstances that we may wish to simulate. At this point, we now look to specific situations and see how the system responds.

For the purposes of this example, we will assume that the slider/pendulum system is initially at rest (all velocity integrators have zero initial conditions). Figure 5-8 shows a sketch of the initial configuration for the example. Note that the center of the slider (denoted by x_s, y_s) is directly above the origin at a height of 1 m. The pendulum is at a 45 ° angle from the vertical.

Tables 5-1 and 5-2 summarize the geometric and inertial parameters, as well as the eight initial conditions that are required by the simulation.

The simulation is then run, and the block is allowed to slide down the ramp. Figure 5-9 shows a plot in which the center of the slider is tracked as well as the pendulum bob.

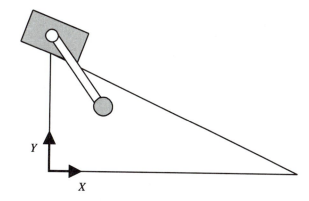

FIGURE 5-8 Initial configuration for the slider/pendulum system.

TABLE 5-1 Inertial and Geometric Parameters

Parameter	Value	Parameter	Value
M_s	5.0 kg	γ	30°
M_p	0.5 kg	r_p	0.5 m
I_p	0.01kg m²		

TABLE 5-2 Initial Conditions for Slider/Pendulum Simulation

Variable	IC	Variable	IC
$\dot{x}_s$	0	x_s	0
$\dot{y}_s$	0	y_s	1.0
$\dot{x}_p$	0	x_p	0.3536
$\dot{y}_p$	0	y_p	0.6464
$\dot{\theta}$	0	θ	$-\pi/4$

Note that the system behaves in a complex manner, yet the simulation's prediction satisfies our intuition. The slider moves down the ramp in a straight line while the pendulum oscillates and thus affects the motion of the slider. Figure 5-10 shows an image of a `Scope` block that is monitoring the normal force between the ramp and the block.

Also, note that the normal force fluctuates over time in response to the swinging of the pendulum.

5.7 Summary

In this chapter, we introduce the simultaneous constraint method by way of a simple example. In the next chapter, the method will be introduced a bit more formally and

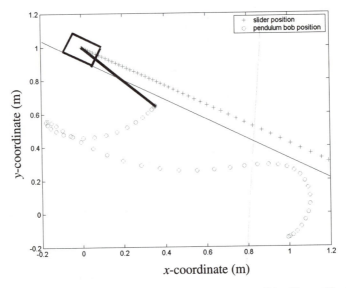

FIGURE 5-9 *x-y* plot of the coordinates of the center of the slider (denoted by "+" signs) and the pendulum bob (shown as "o").

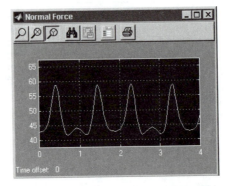

FIGURE 5-10 Normal force between the slider and the block.

applied to more complex systems. The method allows for full dynamic simulation of constrained mechanical systems and the direct computation of the forces of constraint in the system.

CHAPTER 5 PROBLEMS

1. Consider a simple pendulum attached to a horizontal slider as shown in the following figure. The spring is at free length when the *x*-coordinate of the slider is zero.

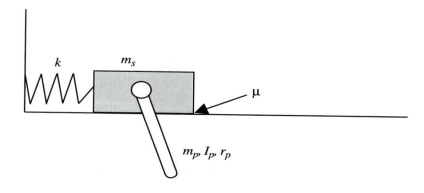

For the values of the parameters listed in the table, plot the response of the slider and the pendulum for various initial positions of the slider. Repeat this for various values of the coefficient of friction, μ.

Parameter	Value	Parameter	Value
k	200 kN/m	m_s	2.2 kg
m_p	0.8 kg	r_p	0.5 m
I_p	0.001 kg m2	m	0.2

2. Consider a flywheel that rotates about a vertical axis. Attached to the flywheel, by means of a link of negligible mass, is a small weight. Construct a dynamic simulation of this system, and graph the angle of the link relative to a radial line on the flywheel as the flywheel starts from rest and experiences a torque of 10 Nm for 20 seconds. Also plot the reaction forces experienced by the flywheel bearings for these conditions. Use the parameter values listed in the table.

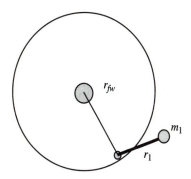

Parameter	Value	Parameter	Value
r_{fw}	25 cm	I_{fw}	0.1 kg m^2
r_1	20 cm	m_1	0.1 kg

3. Consider the flywheel shown below. A small slider is free to move in a slot that has been cut into the flywheel. The slider pushes against a spring, and the spring is at free length when the system is at rest.

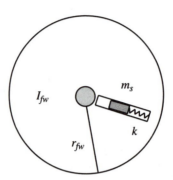

Find the relative deflection of the mass in the slot as the flywheel spins up from rest at a constant acceleration of 10 rad/s². Assume the flywheel has been balanced; use the parameters in the table for your simulation.

Parameter	Value
I_{fw}	1.0 kg m²
r_{fw}	2.0 m
m_s	0.1 kg
k	10000 N/m

CHAPTER 6

THE SIMULTANEOUS CONSTRAINT METHOD

6.1 Introduction

Up to this point in the text, there has been no connection among applied forces, forces within the joints, and the resulting motion of the mechanism. In fact, most kinematic texts seem purposely to obscure that connection. Yet one of the most basic premises of physics for the engineer is that force causes motion. How can so much consideration of the system's motion be possible with no explicit treatment of the forces? The answer lies in the fact that a mechanism is a constrained system, so the nature of relative motions is predetermined by the geometric layout of the device. Also, there are almost always unstated assumptions in most kinematic analyses. For example, when one performs an acceleration and velocity analysis on a mechanism whose input link is moving at a constant speed (see Chapter 4), it is assumed that there is some external device (such as an electric motor) that is providing whatever time-varying torque is required to maintain that speed. Using traditional vector polygon methods, it is extremely tedious to divine that required torque, and it is easy to lose track of the fact that forces, after all, cause the motion being analyzed. In this chapter, work from Chapters 2 and 4 is brought together in a way that allows the formulation of a full dynamic simulation of mechanisms. As before, the slider crank is used as an example. The subsequent chapters provide case studies of other mechanisms to further illustrate the method.

6.2 Description of the Approach

The approach, which is termed the *simultaneous constraint method*, has its roots in many different areas of mechanisms research. Most notable are certain contributions to the robotics field and the work done by E. J. Haug, who has extended this approach to the general, three-dimensional case.[1] Haug's approach is much more general than the one presented here, but it is generally considered graduate-level material due to the complexity of notation required for full spatial representation.

The simultaneous constraint method builds on the kinematic simulation described in Chapter 4. The second derivatives of the components of the loop equation form the

---

[1]Haug, E. J., *Computer-Aided Kinematics and Dynamics of Mechanical Systems*, Englewood Cliffs, NJ: Prentice Hall, 1989.

foundation of the simulation. Then simple force balances are applied to each link to re-late the forces on each link to its individual acceleration. Next, an approach similar to the vector loop approach yields additional information on the accelerations of the link centers of mass. Finally, all the equations are assembled into a sparse matrix that can be solved within MATLAB as part of a full dynamic simulation of the mechanism.

Each step of the approach is now outlined in more detail.

6.2.1 Force Equation

The simultaneous constraint method begins with a free-body diagram of each indi-vidual link. Figure 6-1 shows a typical link as might be found in a mechanism. Note that each joint can sustain only a force (no moment) and that the forces are broken down into x- and y-components. Note also the convention that will be observed when referring to forces that act between links in a mechanism. In general, $\mathbf{F}_{ij}$ will be for force acting on link j, originating from link i. It is easier to remember this convention if you remember that $\mathbf{F}_{ij}$ is the "force of link i acting on link j."

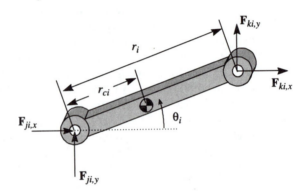

FIGURE 6-1 Free-body diagram of typical link, showing forces due to the revolute joint at each end.

An essential point here is that $\mathbf{F}_{ij}$ is equal and opposite to $\mathbf{F}_{ji}$. This fact will be used implicitly within the method to reduce the number of unknowns in the system of equa-tions.

Referring again to Figure 6-1, the equations of motion for this link can be written:

$$F_{ji,x} + F_{ki,x} = M_i A_{ci,x} \tag{6-1}$$

$$F_{ji,y} + F_{ki,y} = M_i A_{ci,y} \tag{6-2}$$

Equations (6-1) and (6-2) are simply the result of Newton's Second Law of Motion, ap-plied to the link shown in Figure 6-1. The acceleration terms, $A_{ci,x}$ and $A_{ci,y}$ are the x- and y-components of the acceleration of the center of mass of the link, which will be considered shortly.

One last dynamic equation can be written from this free-body diagram: the sum-mation of moments. The general form of the relationship between the summation of moments and the angular acceleration can be rather complicated, but it can be greatly simplified by the proper choice of the axis about which the moments are taken. The two most common choices are any axis that is fixed or an axis that passes through the center of mass (COM) of the body. If one of the two joints are connected with the fixed link, then an axis through that point is preferable. This choice results in a simpler expression because it eliminates the joint forces at that joint. However, if niether joint is connected to the ground link, then the more general expression must be used and we must choose an axis that passes through the center of mass of the object, as shown in Figure 6-2.

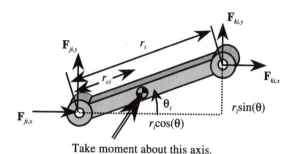

Take moment about this axis.

FIGURE 6-2   Moment arms for the summation of moments equation.

In this case, the summation of moments leads to the following equation:

$$F_{ij,x} r_{ci} \sin(\theta_i) - F_{ij,y} r_{ci} \cos(\theta_i) - F_{ki,x}(r_i - r_{ci})\sin(\theta_i) + F_{ki,y}(r_i - r_{ci})\cos(\theta_i) = I_i \alpha_i \quad (6\text{-}3)$$

Note that I_i is the moment of inertia about the COM of the link. Often the moment is known with respect to a different axis. If that is the case, then the parallel axis theorem may be used to compute the appropriate value.

$$I_{i,\text{new}} = I_i + M_i d^2 \tag{6-4}$$

where d is the distance between the axis that passes through the center of mass and the axis about which I_i is known.

6.2.2   Vector Loop Equations

The vector loop equations have been described in previous chapters. Since the dynamic simulation will be relating forces to accelerations, then the second derivative of the vector loop equations must be taken. Refer to the table at the end of Chapter 2 for the acceleration equations for both the four-bar and the slider crank. These two equations express the relationships that are inherent between the accelerations of the links due to the kinematic constraints imposed by the mechanism.

6.2.3   Vector Equations for COM Accelerations

Before the dynamic equations of motion can be combined with the vector loop equations (also known as constraint equations), one more piece of information must be supplied. Note that while deriving the equations of motion for an individual link, a new

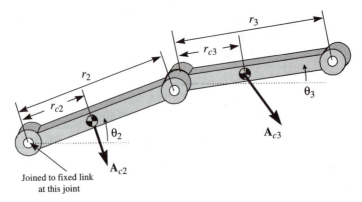

FIGURE 6-3 Portion of linkage to demonstrate method of deriving expression for acceleration of center of mass.

motion variable was introduced, the linear acceleration of the center of mass (COM). In general, those accelerations don't appear in the vector loop equations for the entire mechanism. Therefore, for each link another vector loop equation must be derived that will relate the acceleration of the COM of each link to the other motion variables. Figure 6-3 shows two links that make up a portion of an existing mechanism. To derive the equations for the COM of the two links, start by writing the simple vector relationships by inspection:

$$\mathbf{A}_{c1} = \ddot{\mathbf{R}}_{c1} \tag{6-5}$$

$$\mathbf{A}_{c2} = \ddot{\mathbf{R}}_1 + \ddot{\mathbf{R}}_{c2} \tag{6-6}$$

As with the previous vector equations, these equations can be represented as sets of scalar equations in the x- and y-directions. It is left as an exercise for the student to show that the following equations are equivalent to the previous equations.

$$A_{c2,x} = -r_{c2}\alpha_2 \sin\theta_2 - r_{c2}\omega_2^2 \cos\theta_2 \tag{6-7}$$

$$A_{c2,y} = r_{c2}\alpha_2 \cos\theta_2 - r_{c2}\omega_2^2 \sin\theta_2 \tag{6-8}$$

$$A_{c3,x} = -r_2\alpha_2 \sin\theta_2 - r_2\omega_2^2 \cos\theta_2 - r_{c3}\alpha_3 \sin\theta_3 - r_{c3}\omega_3^2 \cos\theta_3 \tag{6-9}$$

$$A_{c3,y} = r_2\alpha_2 \cos\theta_2 - r_2\omega_2^2 \sin\theta_2 + r_{c3}\alpha_3 \cos\theta_3 - r_{c3}\omega_3^2 \sin\theta_3 \tag{6-10}$$

6.2.4 Implementation of the Dynamic Simulation

The equations discussed in the previous sections—those that result from the application of Newton-Euler laws to each link, those that result from the second derivatives of the original vector loop equations, and those that result from the center-of-mass accelerations—form a set of loosely coupled equations that are linear in forces and accelerations. In general, for an n-link mechanism, there will be $3(n-1)$ Newton-Euler equations, there will be the two component equations from the vector loop, and there will be $2(n-1)$ equations for the centers of mass. Therefore, for a four-bar linkage, this analysis will lead to 17 equations, with 17 unknowns. Note, however, that it is rarely necessary to utilize the most general form of this analysis. Often the COM can be considered to coincide with one of the joint centers (for example, when a crank is actually part of a balanced flywheel, the COM is stationary and those accelerations are zero).

The next section will demonstrate the method for a slider crank.

6.3 Application of Simultaneous Constraint Method for the Slider Crank

By way of illustrating the simultaneous constraint method for dynamic simulations, a full dynamic simulation of a slider crank will be presented in this section. Figure 6-4 shows a slider crank mechanism with two external loads. An external torque, τ_{12} is applied to the crank while a force, $\mathbf{F}_{ext}$, is seen at the slider. In the following sections, the simulation will be assembled for a constant crank speed and constant external force. In this case, the crank speed, ω_2, is considered the input while τ_{12} will become an output. This will allow for the computation of the torque required to maintain that speed under the influence of the external load. Later, the crank speed is considered variable and the torque becomes an input to the simulation.

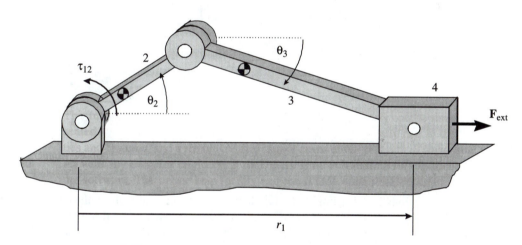

FIGURE 6-4   Schematic of a slider crank with torque input, an externally applied load, and constant input speed.

6.3.1 The Force Equations

As discussed above, the first step is to apply Newton's law of motion to each link. This is aided by free-body diagrams as shown in Figures 6-5 and 6-6.

The equations of motion for link 2 are easily written as shown below.

$$F_{12,x} + F_{32,x} = M_2 A_{c2,x} \tag{6-11}$$

$$F_{12,y} + F_{32,y} = M_2 A_{c2,y} \tag{6-12}$$

$$-F_{32,x} r_2 \sin(\theta_2) + F_{32,y} r_2 \cos(\theta_2) + \tau_{12} = I_{2o} \alpha_2 \tag{6-13}$$

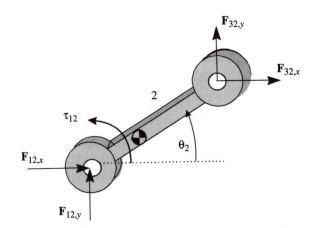

FIGURE 6-5 Free-body diagram of link 2 of the slider-crank mechanism.

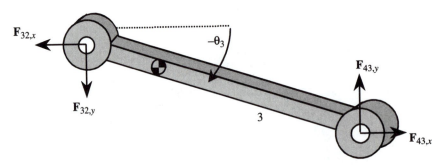

FIGURE 6-6 Free-body diagram of link 3 of the slider-crank mechanism.

Similarly, for link 3, the free-body diagram leads directly to the equations of motion, but with a significant difference. It was noted earlier that the convention by which the forces were named leads to easy identity of equal and opposite forces ($\mathbf{F}_{ij}$ is equal and opposite to $\mathbf{F}_{ji}$). Note that this relationship is used implicitly in Figure 6-6 and in the equations that follow. Instead of introducing new unknown forces ($\mathbf{F}_{23,x}$ and $\mathbf{F}_{23,y}$), the forces that were already introduced ($\mathbf{F}_{32,x}$ and $\mathbf{F}_{32,y}$) are drawn in the free-body diagram in the negative direction and used in the force and moment equations with negative signs. The observant student will note that the resulting equations are identical to those that would have been produced if the new force components were introduced and then substituted by the other forces, making use of the force identity noted above.

$$-F_{32,x} + F_{43,x} = M_3 A_{c3,x} \tag{6-14}$$

$$-F_{32,y} + F_{43,y} = M_3 A_{c3,y} \tag{6-15}$$

$$\begin{aligned}-F_{43,x}(r_3 - r_{c3})\sin\theta_3 + F_{43,y}(r_3 - r_{c3})\cos\theta_3 \\ - F_{32,x}r_{c3}\cos\theta_3 + F_{32,y}r_{c3}\sin\theta_3 = I_3\alpha_3\end{aligned} \tag{6-16}$$

Finally, link 4, the slider block, is shown with all forces acting upon it (Figure 6-7).

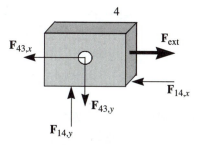

FIGURE 6-7 Free-body diagram of the slider block.

Note that there is no moment equation for link 4, since the slider block exists as a geometric point as far as the kinematic analysis is concerned. Note also that force component $F_{14,x}$ is the friction encountered by the slider. The present analysis considers a frictionless interface between the slider and the fixed link, but the force is shown in the figure for the sake of completeness and to guide the student for more advanced analysis. Finally note that the prismatic joint between the slider and the fixed link constrains the block to move only in the x-direction. Therefore the y-component of the acceleration is zero.

$$F_{34,x} + F_{ext} = M_4 \ddot{r}_1 \tag{6-17}$$

$$F_{34,y} + F_{14,y} = 0 \tag{6-18}$$

Now, consider the unknowns. Through examination of equations (6-4) through (6-18), the following forces are unknown:

$$\begin{bmatrix} F_{12,x} & F_{12,y} & F_{32,x} & F_{32,y} & F_{34,x} & F_{34,y} & F_{14,y} & \tau_{12} \end{bmatrix}$$

In addition, the following accelerations are introduced, and additional equations must be derived from the vector loop equations to complete our set of equations:

$$\begin{bmatrix} A_{c2,x} & A_{c2,y} & A_{c3,x} & A_{c3,y} & \ddot{r}_1 & \alpha_2 & \alpha_3 \end{bmatrix}$$

6.3.2 The Vector Loop Equations

The vector loop equations for a slider crank have been derived previously and are repeated here for the sake of completeness. The derivation of these equations can be found in Chapter 4.

$$\ddot{r}_1 + r_3 \sin\theta_3 \alpha_3 + r_2 \sin\theta_2 \alpha_2 = -r_2 \cos\theta_2 \omega_2^2 - r_3 \cos\theta_3 \omega_3^2 \tag{6-19}$$

$$-r_3 \cos\theta_3 \alpha_3 + r_2 \cos\theta_2 \alpha_2 = -r_2 \sin\theta_2 \omega_2^2 - r_3 \sin\theta_3 \omega_3^2 \tag{6-20}$$

6.3.3 Center-of-Mass Accelerations

Equations (6-7) through (6-10) complete the set of equations required to simulate the slider crank. These last four equations relate the kinematic state of the mechanism (link displacement, velocity, and accelerations) with the acceleration components of the centers of mass of the links, which are required in the force equations. They are repeated as follows:

$$A_{c2,x} = -r_{c2}\alpha_2 \sin\theta_2 - r_{c2}\omega_2^2 \cos\theta_2 \tag{6-21}$$

$$A_{c2,y} = r_{c2}\alpha_2 \cos\theta_2 - r_{c2}\omega_2^2 \sin\theta_2 \tag{6-22}$$

$$A_{c3,x} = -r_2\alpha_2 \sin\theta_2 - r_2\omega_2^2 \cos\theta_2 - r_{c3}\alpha_3 \sin\theta_3 - r_{c3}\omega_3^2 \cos\theta_3 \tag{6-23}$$

$$A_{c3,y} = r_2\alpha_2 \cos\theta_2 - r_2\omega_2^2 \sin\theta_2 + r_{c3}\alpha_3 \cos\theta_3 - r_{c3}\omega_3^2 \sin\theta_3 \tag{6-24}$$

6.3.4 Assembling the System of Equations

The final step before building the Simulink simulation is to assemble all 14 equations in matrix form. This requires that the inputs of the system be identified, since only known quantities (inputs and link velocities) will be allowed on the righthand side of the equation. Recall at the beginning of the section, it was stated that the first simulation will be the constant-speed slider crank. Therefore, the torque applied to the crank is an unknown and α_2 is zero. The complete set of equations in vector form is shown below:

$$
\begin{bmatrix}
1 & 0 & 1 & 0 & 0 & 0 & 0 & 0 & 0 & 0 & -M_2 & 0 & 0 & 0 \\
0 & 1 & 0 & 1 & 0 & 0 & 0 & 0 & 0 & 0 & 0 & -M_2 & 0 & 0 \\
0 & 0 & -r_2S_2 & r_2C_2 & 0 & 0 & 0 & 1 & 0 & 0 & 0 & 0 & 0 & 0 \\
0 & 0 & -1 & 0 & 1 & 0 & 0 & 0 & 0 & 0 & 0 & 0 & -M_3 & 0 \\
0 & 0 & 0 & -1 & 0 & 1 & 0 & 0 & 0 & 0 & 0 & 0 & 0 & -M_3 \\
0 & 0 & -r_{c3}C_3 & r_{c3}S_3 & (r_{c3}-r_3)S_3 & (r_3-r_{c3})C_3 & 0 & 0 & 0 & -I_3 & 0 & 0 & 0 & 0 \\
0 & 0 & 0 & 0 & 1 & 0 & 0 & 0 & -M_4 & 0 & 0 & 0 & 0 & 0 \\
0 & 0 & 0 & 0 & 0 & 1 & 1 & 0 & 0 & 0 & 0 & 0 & 0 & 0 \\
0 & 0 & 0 & 0 & 0 & 0 & 0 & 0 & 1 & r_3S_3 & 0 & 0 & 0 & 0 \\
0 & 0 & 0 & 0 & 0 & 0 & 0 & 0 & 0 & -r_3C_3 & 0 & 0 & 0 & 0 \\
0 & 0 & 0 & 0 & 0 & 0 & 0 & 0 & 0 & 0 & 1 & 0 & 0 & 0 \\
0 & 0 & 0 & 0 & 0 & 0 & 0 & 0 & 0 & 0 & 0 & 1 & 0 & 0 \\
0 & 0 & 0 & 0 & 0 & 0 & 0 & 0 & 0 & r_{c3}S_3 & 0 & 0 & 1 & 0 \\
0 & 0 & 0 & 0 & 0 & 0 & 0 & 0 & 0 & -r_{c3}C_3 & 0 & 0 & 0 & 1 \\
\end{bmatrix}
\begin{bmatrix}
F_{12,x} \\
F_{12,y} \\
F_{32,x} \\
F_{32,y} \\
F_{43,x} \\
F_{43,y} \\
F_{14,y} \\
\tau_{12} \\
\ddot{r}_1 \\
\alpha_3 \\
A_{c2,x} \\
A_{c2,y} \\
A_{c3,x} \\
A_{c3,y} \\
\end{bmatrix}
$$

$$
=
\begin{bmatrix}
0 \\
0 \\
0 \\
0 \\
0 \\
0 \\
0 \\
-F_{ext} \\
0 \\
-r_2C_2\omega_2^2 - r_3C_3\omega_3^2 \\
-r_2S_2\omega_2^2 - r_3S_3\omega_3^2 \\
-r_{c2}C_2\omega_2^2 \\
-r_{c2}S_2\omega_2^2 \\
-r_2C_2\omega_2^2 - r_{c3}C_3\omega_3^2 \\
-r_2S_2\omega_2^2 - r_{c3}S_3\omega_3^2 \\
\end{bmatrix}
\tag{6-25}
$$

6.4 Dynamic Simulation of the Slider Crank

Using the same approach outlined in Chapter 4 for kinematic simulations, we assemble the full dynamic simulation of the mechanism. For illustrative purposes, we choose the situation in which the input link (the crank) is assumed to turn at a constant rate of 100 rad/s. In this case, five integrators will be used: one to integrate the crank velocity to produce the crank angle (θ_2); two more integrators to integrate the accelerations solved by the constraint equations (θ_3 and $\ddot{r}_1$); and, finally, two more to integrate those resulting velocities to produce displacements.

The matrix equation derived in the previous section (equation 6-25) is solved using an m-file function that will take all of the integrator outputs as arguments. In addition, we allow the possibility of an external force on the slider. Finally, the m-file will also perform a consistency check to assure us that the formulation is error-free and that the integration routines are maintaining adequate accuracy.

Figure 6-8 shows one possible realization of the simulation in Simulink.

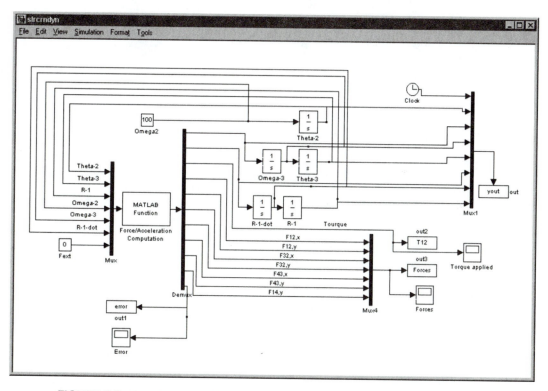

FIGURE 6-8 Simulink realization of full dynamic simulation of a slider-crank mechanism (file: `slicrndyn.mdl`).

The function to compute the acceleration and forces is shown in the following listing. Note that all of the mechanism parameters must be defined within this function. The link lengths, r_2 and r_3, are the obvious parameters, but the inertial properties of each link, the mass, the moment of inertia about COM, and the location of the COM relative to the link coordinate frame must also be specified. For this example, we take a set of parameters that might be found in a typical single-cylinder lawn mower engine.

---

slrcrndy.m

```
function [xout]=slrcrndy(u)
%
%   Function to implement the full dynamic simulation
%   of a slider crank.
%
% Used with SLRCRNDYN.MDL SIMULINK file
%
%
%   BY:  J.F. Gardner
%
%
% u(1) = Theta-2
% u(2) = Theta-3
% u(3) = r-1
% u(4) = Omega-2
% u(5) = Omega-3
% u(6) = r-1-dot
% u(7) = F-ext
%
% Define some local variables
%
r1   = u(3);
r2   = 0.05; % crank length in meters (about 2 inches)
r3   = 0.25; % connecting rod length in meters
rc2  = 0.0;  % COM at pivot, implies balanced crank
rc3  = r3/3; % COM about 1/3 of the distance between big end and wrist
pin
C2   = cos(u(1));  S2   = sin(u(1));
C3   = cos(u(2));  S3   = sin(u(2));
w2   = u(4);  w3   = u(5);
Fext = u(7);
%
%   Define inertial properties
%
M2 = 1.0; % Mass of crank in kg
M3 = 0.2; % Mass of conn. rod in kg
M4 = 0.2; % Mass of piston in kg
I30 = .001; % Mass moment of conn rod about COM in kg-m^2
%
a=zeros(14);
b=zeros(14,1);
%
a(1,1) = 1;  a(1,3) = 1;  a(1,11) = -M2;
a(2,2) = 1;  a(2,4) = 1;  a(2,12) = -M2;
a(3,3) = -r2*S2;  a(3,4) =  r2*C2;  a(3,8) =  1;
a(4,3) = -1;  a(4,5) =  1;  a(4,13) = -M3;
```

```
a(5,4)  = -1;   a(5,6)  =   1;   a(5,14) = -M3;
a(6,5)  = r3*S3;  a(6,6) = -r3*C3;  a(6,10) = -I30;
a(7,5)  = 1;   a(7,9)  = -M4;
a(8,6)  = 1;   a(8,7)  = 1;
a(9,9)  = 1;   a(9,10) = r3*S3;
a(10,10) = -r3*C3;
a(11,11) = 1;
a(12,12) = 1;
a(13,10) = rc3*S3;  a(13,13) = 1;
a(14,10) = -rc3*C3;  a(14,14) = 1;
%
%  Set up the RHS vector of equation (5-25)
b(7)  = -Fext;
b(9)  = -r2*C2*w2^2-r3*C3*w3^2;
b(10) = -r2*S2*w2^2-r3*S3*w3^2;
b(11) = -rc2*C2*w2^2;
b(12) = -rc2*S2*w2^2;
b(13) = -r2*C2*w2^2-rc3*C3*w3^2;
b(14) = -r2*S2*w2^2-rc3*S3*w3^2;
%
%  Solve the equations
%
x = inv(a)*b;
%
%  Compute the consistency error
%
error = norm([r1-r2*C2-r3*C3,r2*S2+r3*S3]);
%
%  Set up output vector
%
xout(1)  = x(10);  % Alpha-3
xout(2)  = x(9);   % r1-double-dot
xout(3)  = x(8);   % Torque
xout(4)  = x(1);   % F12x
xout(5)  = x(2);   % F12y
xout(6)  = x(3);   % F32x
xout(7)  = x(4);   % F32y
xout(8)  = x(5);   % F43x
xout(9)  = x(6);   % F43y
xout(10) =x(7);    % F14y
xout(11) = error;  % Consistency Error
```

The remaining step before the simulation can be run is determining the initial conditions for the five integrators. As was the case for the kinematic simulation, a consistent set of initial conditions is essential. If we choose to begin at $\theta_2 = 0$, then the initial positions are simple ($\theta_3 = 0.0$ and $r_1 = r_2 + r_3 = 0.5$ m). The velocities are not quite so easy as they must represent a viable solution of the mechanism when the crank is moving at 100 rad/s and is at zero degrees. Referring to the velocity equations for the inline slider crank shown in equation (4-6), we can use MATLAB at the command prompt to compute the solution to a high degree of accuracy:

```
» r2=0.05;
» r3=0.25;
» om2=100;
» a=[r3*sin(0)   1; -r3*cos(0)   0]
a =
            0    1.0000
     -0.2500         0
» b = [-r2*om2*sin(0); r2*om2*cos(0)];
» vels=inv(a)*b
vels =
      -20
        0
```

This result implies that the initial condition for ω_3 is –20 rad/s and for $\dot{r}_1$ is 0 (consistent with the fact that the initial position is top dead center for the slider crank).

Before we move on, it's important to point out that the process of computing initial conditions is greatly simplified by the clever choice of initial position of the mechanism. In the case of the present example, top dead center (TDC) for a slider crank represents an excellent choice since the crank and connecting rod (r_2 and r_3) are both at zero angles (assuming that the x-axis is aligned with the slider axis) and the velocity of the piston is also zero. Therefore, the only value that must be computed is ω_3. While using MATLAB to solve the velocity equation for this pose (as shown above) is a viable option, the observant engineer will note that the equations are decoupled and simplified for this particular pose. In fact, we can solve for ω_3 in closed form as shown below:

$$\omega_3 = -\frac{r_2}{r_3}\omega_2$$

6.5 Simulation Studies of the Slider Crank

A full dynamic simulation of a mechanism is capable of delivering far more information than a simple kinematic solution. In particular, the dynamic forces that arise due to the accelerating inertia in the mechanism are computed and easily found.

We choose to run the simulation developed in the previous section for a period of 0.1 second, which corresponds to slightly less than two complete rotations of the crank. It's important to note that for this situation, only one complete rotation of the crank need be simulated since the simulation is built on the assumption that the crank moves at a constant rotational rate. The simulation will uncover no different behavior for the mechanism after the first rotation.

Figure 6-9 describes the torque that must be delivered to the slider crank in order to maintain constant speed. Further interpretation of this plot is in order. The analysis began with the assumption that the crank was moving with a constant angular rate of 100 rad/s (approximately 950 rpm), but there was no mention of the device that was causing this motion to occur. Note that the torque is both positive and negative, implying that power must be delivered to and removed from the mechanism over the course of one rotation. This is consistent with the fact that we modeled no external load and no inter-

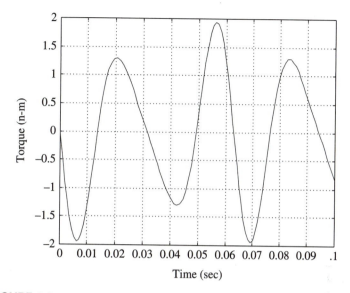

FIGURE 6-9 Torque required to maintain a constant crank speed of 100 rad/s.

nal friction to dissipate energy. Therefore, the net energy transferred over time must average out to zero to satisfy the conservation of energy.

In applications, the assumption of a constant crank speed must be carefully considered. If the input member has a rather large rotary inertia and the prime mover has a torque capacity that is large relative to the demands of the job, then the assumption is usually warranted. However, it is often necessary that we extend our dynamic model to include a simple representation of the motor that is driving the device so that the speed fluctuations of the crank may be considered.

In addition to the input torque, this simulation can also compute the forces of constraint within the mechanism. For example, $F_{12,x}$ and $F_{12,y}$ are the x- and y-components of the reaction force acting on the crank hub at the junction with the fixed link. This force is an important variable to consider if the bearings of this device are to be designed. In fact, the dynamic forces predicted in this simulation are the only bearing forces, since there are no external forces acting on this device.

Figure 6-10 is a plot of the x- and y-components of this force for the same simulation run. Note that this figure can be difficult to interpret as it is not clear by inspection how the force magnitude behaves over time. While the combination of orthogonal force components to yield magnitude and direction is a trivial task, Simulink can be used to make the job even easier.

Figure 6-11 shows a segment of the Simulink block diagram that can used to perform this task. Note that the x- and y-components can be considered analogous to rectangular coordinates, and the desired magnitude and direction values are the corresponding polar coordinates. Simulink has predefined blocks that can be used to perform this transformation. Figure 6-11 shows a fragment of the Simulink model that has been expanded to include conversion blocks. These blocks are intended to be used to convert rectangular coordinates to polar coordinates and are perfectly suited to the

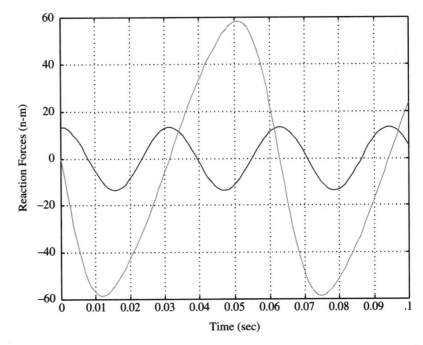

FIGURE 6-10 *x*- and *y*-components of the ground reaction force at the main crank bearing.

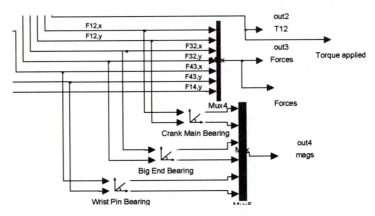

FIGURE 6-11 Section of Simulink diagram incorporating blocks used
to transform rectangular coordinates to polar coordinates.

task of computing force magnitude and direction, given the *x*- and *y*-components. This
block is found in the "Simulink Extras" library and is called "Cartesian to Polar."

Figure 6-12 shows a plot of the magnitudes of forces F_{12}, F_{23}, and F_{34} over time
while Figure 6-13 shows the force direction measured from the positive *x*-axis.

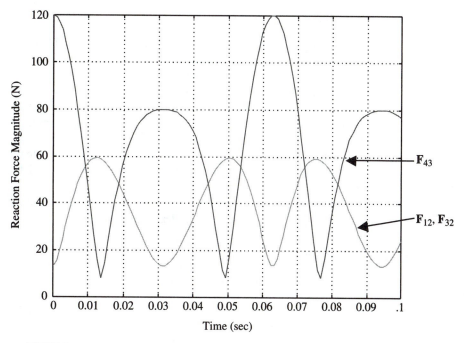

FIGURE 6-12 Magnitude of crank main bearing force, connecting rod "big end" force and wrist pin force for the slider-crank mechanism under consideration.

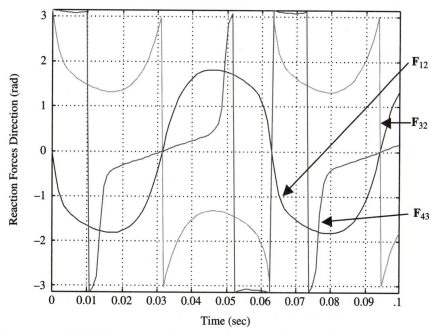

FIGURE 6-13 Direction of the three forces for the slider crank.

6.6 Summary

In this chapter, we present the simultaneous constraint method that is a powerful technique for developing fully dynamic simulation of constrained systems. The major advantage of a dynamic simulation over a kinematically driven dynamic analysis is that no assumptions about the motion of the system are required. For example, the vast majority of dynamic analyses of mechanisms begin with an assumption that one of the links is moving at a constant speed. In reality there is always interplay between the system that is acting as the prime mover (for example, an electric motor) and the inertial loads imposed by a mechanism. As the torque demand increases, the motor will tend to slow down. The only way to see this effect is with a dynamic simulation such as those outlined in this chapter.

In subsequent chapters, we will examine some variations on the theme of constrained systems and look at two examples in greater detail.

CHAPTER 6 PROBLEMS

1. Consider the double-slider mechanism shown below.

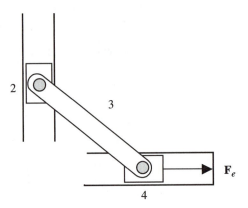

Construct a dynamic simulation in which the input is the external force, $\mathbf{F}_e$, as shown. Using the parameters from the table below, plot the normal forces on the slides for $\mathbf{F}_e = 10 \sin(5t)$.

Parameter	Value	Parameter	Value
r_3	1.0 m	m_3	1.2 kg
m_2	0.8 kg	m_4	1.5 kg
I_3	0.01 kg m2		

2. A compound pendulum is one in which two or more pendula are connected in series. These are fascinating mechanisms in that they exhibit chaotic behavior. Among other things, chaotic systems are ones that are extremely sensitive to initial conditions. Two sets of initial conditions that are very close to one another can

bring about very different behaviors of the system. Construct a dynamic simulation of a double pendulum with the following parameter values.

Parameters	Value
r_1	10 cm
r_{c1}	4 cm
r_2	8 cm
r_{c2}	5 cm
m_1	0.1 kg
I_1	0.001 kg m^2
m_2	0.08 kg
I_2	0.0006 kg m^2

Use the simulation to explore the sensitivity of the system response to small changes in the initial conditions.

3. Modify the slider-crank simulation shown in the text to accommodate offset slider cranks. Using the inertial and geometric parameters from the text example, explore the effect of the offset on the normal forces on the slider. Make a plot of maximum normal force at 1800 rpm versus offset from –0.02 to 0.02 m, in increments of 0.005 m.

CHAPTER 7

TWO-LINK PLANAR ROBOT

7.1 Overview

In this chapter, we consider the simulation of an open-chain mechanism that is commonly encountered in the robotics literature. The two-link planar robot is a simple, two-degree-of-freedom mechanism that incorporates a surprising amount of complexity in its dynamics. Figure 7-1 shows the pertinent geometric parameters for this mechanism.

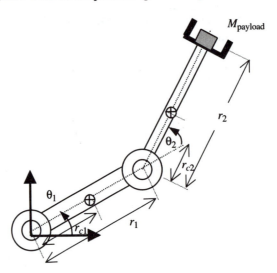

FIGURE 7-1 The two-link robot mechanism.

In this chapter, we will derive the appropriate vector equations and dynamic equations for this mechanism. The resulting dynamic simulation will have two inputs—the torques generated by the two motors.

7.2 Vector Equations

The vector equation for the two-link robot is easily derived:

$$\mathbf{R}_{pl} = \mathbf{R}_1 + \mathbf{R}_2 \tag{7-1}$$

The corresponding x- and y-components are shown below. Note that the form of these equations is slightly different from those we've seen thus far. The reason for these differences lies in the manner in which we define the vector angles. In Figure 7-1, the angles are referenced to the orientation of the previous link, not the global x-axis. This convention is common in robotics and arises from the fact that sensors on robot arms measure the relative angles, not the absolute values.

$$x_{pl} = r_1 \cos\theta_1 + r_2 \cos(\theta_1 + \theta_2) \tag{7-2}$$

$$y_{pl} = r_1 \sin\theta_1 + r_2 \sin(\theta_1 + \theta_2) \tag{7-3}$$

Taking the derivative:

$$\dot{x}_{pl} = -r_1\omega_1 \sin\theta_1 - r_2(\omega_1 + \omega_2)\sin(\theta_1 + \theta_2) \tag{7-4}$$

$$\dot{y}_{pl} = r_1\omega_1 \cos\theta_1 + r_2(\omega_1 + \omega_2)\cos(\theta_1 + \theta_2) \tag{7-5}$$

these can be rearranged to give us the following matrix equation:

$$\begin{bmatrix} \dot{x}_{pl} \\ \dot{y}_{pl} \end{bmatrix} = \begin{bmatrix} -r_1 S_1 - r_2 S_{12} & -r_2 S_{12} \\ r_1 C_1 + r_2 C_{12} & r_2 C_{12} \end{bmatrix} \begin{bmatrix} \omega_1 \\ \omega_2 \end{bmatrix} \tag{7-6}$$

This relationship between the joint rates and the end-effector velocities in Cartesian coordinates is well known in robotics. The matrix is known as the Jacobian matrix, and it can form the basis of a very effective control algorithm.

The derivatives of these equations lead to the following acceleration equations:

$$\ddot{x}_{pl} + (r_1 S_1 + r_2 S_{12})\alpha_1 + r_2 S_{12}\alpha_2 = -[(r_1 C_1 + r_2 C_{12})\omega_1^2 + r_2 C_{12}\omega_2^2 + 2r_2\omega_1\omega_2 C_{12}] \tag{7-7}$$

$$\ddot{y}_{pl} - (r_1 C_1 + r_2 C_{12})\alpha_1 - r_2 C_{12}\alpha_2 = -[(r_1 S_1 + r_2 S_{12})\omega_1^2 + r_2 S_{12}\omega_2^2 + 2r_2\omega_1\omega_2 S_{12}] \tag{7-8}$$

These equations form the basis of our dynamic simulation; they express the relationship between the acceleration of the payload and the angular accelerations and velocities of the two joint motors. As we've seen in previous dynamic simulations, it is also important that we find the relationship between the accelerations of the centers of mass of the two links and the joint variables. Those relationships are easily derived and summarized below:

$$A_{c1,x} + r_{c1} S_1 \alpha_1 = -r_{c1} C_1 \omega_1^2 \tag{7-9}$$

$$A_{c1,y} - r_{c1} C_1 \alpha_1 = -r_{c1} S_1 \omega_1^2 \tag{7-10}$$

$$A_{c2,x} + (r_1 S_1 + r_{c2} S_{12})\alpha_1 + r_{c2} S_{12}\alpha_2 = -[(r_1 C_1 + r_{c2} C_{12})\omega_1^2 + r_{c2} C_{12}\omega_2^2 + 2r_{c2}\omega_1\omega_2 C_{12}] \tag{7-11}$$

$$A_{c2,y} - (r_1 C_1 + r_{c2} C_{12})\alpha_1 - r_{c2} C_{12}\alpha_2 = -[(r_1 S_1 + r_{c2} S_{12})\omega_1^2 + r_{c2} S_{12}\omega_2^2 + 2r_{c2}\omega_1\omega_2 S_{12}] \tag{7-12}$$

7.3 Dynamic Equations

Following the methods outlined in Chapter 6, we will examine the free-body diagrams of each link separately. Figure 7-2 shows the free-body diagram for the first link of the planar robot.

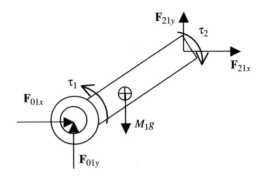

FIGURE 7-2 Free-body diagram of the first link of the
two-link planar robot shown in Figure 7-1.

The three equations of motion that can be derived from this link are shown below:

$$F_{01x} + F_{21x} = M_1 A_{c1,x} \tag{7-13}$$

$$F_{01y} + F_{21y} - M_1 g = M_1 A_{c1,y} \tag{7-14}$$

$$\tau_1 - \tau_2 - F_{21x} r_1 S_1 + F_{21y} r_1 C_1 - M_1 g r_{c1} C_1 = I_1 \alpha_1 \tag{7-15}$$

Similarly, the free-body diagram for the second link is shown in Figure 7-3, and the equations of motion follow.

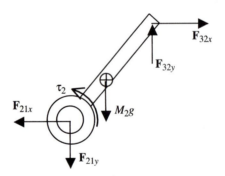

FIGURE 7-3 Free-body diagram of the second link of
the two-link planar robot shown in Figure 7-1.

$$F_{32x} - F_{21x} = M_2 A_{c2,x} \tag{7-16}$$

$$F_{32y} - F_{21y} - M_2 g = M_2 A_{c2,y} \tag{7-17}$$

$$\tau_2 - F_{21x} r_{c2} S_{12} + F_{21y} r_{c2} C_{12} - F_{32x} (r_2 - r_{c2}) S_{12} + F_{32y} (r_2 - r_{c2}) C_{12} = I_2 \alpha_2 \tag{7-18}$$

Finally, we examine the payload of the manipulator. Note that since the motion of the payload is directly related to link 2, we could have lumped it in with that link in Figure 7-3. However, since the mass of the payload may change as the robot picks up different objects, and we may be interested in the forces required to hold on to that payload, we find it more convenient to add a few more equations of motion so that we can track its motion separately.

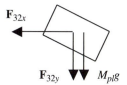

FIGURE 7-4 Free-body diagram of the payload.

We consider the payload as a lumped mass, and therefore it yields only two equations.

$$M_{pl}\ddot{x}_{pl} = -F_{32x} \tag{7-19}$$

$$M_{pl}\ddot{y}_{pl} = -F_{32y} - M_{pl}g \tag{7-20}$$

Summarizing, we find six equations from the vector equations and eight from the dynamic equations. If the motor torques are considered inputs to the system, then we have the following 14 unknowns:

$$[\alpha_1 \quad \alpha_2 \quad A_{c1x} \quad A_{c1y} \quad A_{c2x} \quad A_{c2y} \quad A_{plx} \quad A_{ply} \quad F_{01x} \quad F_{01y} \quad F_{21x} \quad F_{21y} \quad F_{32x} \quad F_{32y}]$$

7.4 The Simultaneous Constraint Matrix

Combining the six kinematic constraint equations with the eight dynamic equations results in a system of 14 equations that are linear in the 14 unknowns, as demonstrated below:

$$
\begin{bmatrix}
r_1S_1+r_2S_{12} & r_2S_{12} & 0 & 0 & 0 & 0 & 1 & 0 & 0 & 0 & 0 & 0 & 0 & 0 \\
-r_1C_1-r_2C_{12} & -r_2C_{12} & 0 & 0 & 0 & 0 & 0 & 1 & 0 & 0 & 0 & 0 & 0 & 0 \\
r_{c1}S_1 & 0 & 1 & 0 & 0 & 0 & 0 & 0 & 0 & 0 & 0 & 0 & 0 & 0 \\
-r_{c1}C_1 & 0 & 0 & 1 & 0 & 0 & 0 & 0 & 0 & 0 & 0 & 0 & 0 & 0 \\
r_1S_1+r_2S_{12} & r_{c2}S_{12} & 0 & 0 & 1 & 0 & 0 & 0 & 0 & 0 & 0 & 0 & 0 & 0 \\
-r_1C_1-r_2C_{12} & -r_{c2}C_{12} & 0 & 0 & 0 & 1 & 0 & 0 & 0 & 0 & 0 & 0 & 0 & 0 \\
0 & 0 & -M_1 & 0 & 0 & 0 & 0 & 0 & 1 & 0 & 1 & 0 & 0 & 0 \\
0 & 0 & 0 & -M_1 & 0 & 0 & 0 & 0 & 0 & 1 & 0 & 1 & 0 & 0 \\
I_1 & 0 & 0 & 0 & 0 & 0 & 0 & 0 & 0 & 0 & r_1S_1 & -r_1C_1 & 0 & 0 \\
0 & 0 & 0 & 0 & -M_2 & 0 & 0 & 0 & 0 & 0 & -1 & 0 & 1 & 0 \\
0 & 0 & 0 & 0 & 0 & -M_2 & 0 & 0 & 0 & 0 & 0 & -1 & 0 & 1 \\
0 & I_2 & 0 & 0 & 0 & 0 & 0 & 0 & 0 & 0 & r_{c2}S_{12} & -r_{c2}C_{12} & (r_2-r_{c2})S_{12} & -(r_2-r_{c2})C_{12} \\
0 & 0 & 0 & 0 & 0 & 0 & M_{pl} & 0 & 0 & 0 & 0 & 0 & 1 & 0 \\
0 & 0 & 0 & 0 & 0 & 0 & 0 & M_{pl} & 0 & 0 & 0 & 0 & 0 & 1
\end{bmatrix}
\begin{bmatrix}
\alpha_1 \\ \alpha_2 \\ A_{c1x} \\ A_{c1y} \\ A_{c2x} \\ A_{c2y} \\ A_{plx} \\ A_{ply} \\ F_{01x} \\ F_{01y} \\ F_{21x} \\ F_{21y} \\ F_{32x} \\ F_{32y}
\end{bmatrix}
$$

$$
=
\begin{bmatrix}
-[(r_1 C_1 + r_2 C_{12})\omega_1^2 + r_2 C_{12}\omega_2^2 + 2r_2\omega_1\omega_2 C_{12}] \\
-[(r_1 S_1 + r_2 S_{12})\omega_1^2 + r_2 S_{12}\omega_2^2 + 2r_2\omega_1\omega_2 S_{12}] \\
-r_{c1} C_1 \omega_1^2 \\
-r_{c1} S_1 \omega_1^2 \\
-[(r_1 C_1 + r_{c2} C_{12})\omega_1^2 + r_{c2} C_{12}\omega_2^2 + 2r_{c2}\omega_1\omega_2 C_{12}] \\
-[(r_1 S_1 + r_{c2} S_{12})\omega_1^2 + r_{c2} S_{12}\omega_2^2 + 2r_{c2}\omega_1\omega_2 S_{12}] \\
0 \\
M_1 g \\
\tau_1 - \tau_2 - M_1 g\, r_{c1} C_1 \\
0 \\
M_2 g \\
\tau_2 \\
0 \\
-M_{pl} g
\end{bmatrix}
\tag{7-21}
$$

This matrix equation is embedded in a MATLAB function that, in turn can be embedded into a Simulink simulation.

7.5 Dynamic Simulation

The Simulink model is assembled using the method described in Chapter 6. Figure 7-5 shows a skeleton of an appropriate Simulink model. Note that the two motor torques are inputs to the system. Also note that a simple model of the bearing friction has also been added. In general, friction is a force or torque opposing motion and related to velocity. In Figure 7-5, a linear friction model is implemented in which torques proportional to the velocity are subtracted from the input torques. The two gain blocks labeled "Damping" represent velocity-dependent losses due to viscous friction in the bearings and motors. The actual values of these coefficients are difficult to discover; however, some type of energy dissipation is certainly present in manipulators of this sort, and the lack of such dissipation in the model would lead to highly misleading and unrealistic results.

The function that solves the 14×14 matrix equation is listed below:

robot.m

```
function out=robot(u)
%
%   u(1)  = omega-1
%   u(2)  = theta-1
%   u(3)  = omega-2
%   u(4)  = theta-2
%   u(5)  = Torque-1
%   u(6)  = Torque-2
%
%
g=9.8067;
%
r1 = 1.0;   rc1 = 0.5;
r2 = 0.8;   rc2 = 0.1;
```

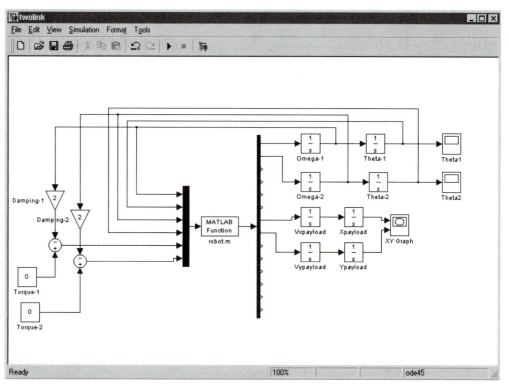

FIGURE 7-5 Simulink model of the two-link robotic manipulator. The outputs of the DeMux block are the 14 unknowns of the matrix equation. The unconnected outputs represent redundant accelerations and the constraint forces (file: twolink.mdl).

```
m1 = 2.5;  m2 = 1.8;
I1 = 0.15;  I2 = 0.05;
%
mpl=2.0;
%
S1 = sin(u(2));  S12 = sin(u(2)+u(4));
C1 = cos(u(2));  C12 = cos(u(2)+u(4));
%
a=zeros(14,14);
b=zeros(14,1);
%
a(1,1) = r1*S1+r2*S12;  a(1,2)=r2*S12; a(1,7) = 1;
a(2,1) = -r1*C1-r2*C12; a(2,2)=-r2*C12; a(2,8) = 1;
a(3,1)=rc1*S1; a(3,3)=1;
a(4,1)=-rc1*C1;a(4,4)=1;
a(5,1)=r1*S1+rc2*S12;a(5,2)=rc2*S12;a(5,5)=1;
a(6,1)=-r1*C1-rc2*C12;a(6,2)=-rc2*C12;a(6,6)=1;
a(7,3)=-m1;a(7,9)=1;a(7,11)=1;
a(8,4)=-m1;a(8,10)=1;a(8,12)=1;
```

```
a(9,1)=I1;a(9,11)=r1*S1;a(9,12)=-r1*C1;
a(10,5)=-m2;a(10,11)=-1;a(10,13)=1;
a(11,6)=-m2;a(11,12)=-1;a(11,14)=1;
a(12,2)=I2;a(12,11)=rc2*S12;a(12,12)=-rc2*C12;a(12,13)=(r2-
rc2)*S12;a(12,14)=-(r2-rc2)*C12;
a(13,7)=mpl;a(13,13)=1;
a(14,8)=mpl;a(14,14)=1;
%
%
b(1)  = -((r1*C1+r2*C12)*u(1)^2+r2*C12*u(3)^2+2*r2*u(1)*u(3)*C12);
b(2)  = -((r1*S1+r2*S12)*u(1)^2+r2*S12*u(3)^2+2*r2*u(1)*u(3)*S12);
b(3)  = -rc1*C1*u(1)^2;
b(4)  = -rc1*S1*u(1)^2;
b(5)  = -((r1*C1+rc2*C12)*u(1)^2+rc2*C12*u(3)^2+2*rc2*u(1)*u(3)*C12);
b(6)  = -((r1*S1+rc2*S12)*u(1)^2+rc2*S12*u(3)^2+2*rc2*u(1)*u(3)*S12);
b(8)  = m1*g;
b(9)  = u(5)-u(6)-m1*g*rc1*C1;
b(11)= m2*g;
b(12)= u(6);
b(14)= -mpl*g;
%
out=inv(a)*b;
```

To gain some confidence in the simulation, a simple test will be performed. Recall from our free-body diagrams that the robot is operating in a vertical plane and that gravity acts on the links. Therefore, if we start the robot from any initial pose, and set the input torques to zero, then the robot should fall under its own weight and settle to a position in which both links extend straight downward. Referring to the definitions of the joint angles in Figure 7-1, this corresponds to joint angles of $\theta_1 = -\pi/2$ and $\theta_2 = 0$.

For initial conditions, we choose $\theta_1 = 0$ and $\theta_2 = \pi/2$ radians. This corresponds to a payload position (the point at the end of the manipulator) of $x_{pl} = 1.0$ and $y_{pl} = 0.8$. Recall that, as was the case with all simulations, the initial conditions of the integrators must be consistent.

Figure 7-6 shows the results of this simulation. Superimposed on the plot are sketches of the manipulator in the initial and final conditions. The data points represent the location of the payload as time progresses. Note that the behavior indicated in the plot is consistent with our intuition of how the robot would behave if allowed to fall under its own weight.

Finally, we'll add a consistency check to further increase our confidence in the simulation. The simulation computes redundant information that should remain consistent if everything is working well. In particular, the x- and y-coordinates of the payload as computed by the integrators should be equal to the x- and y-coordinates computed by using values of θ_1 and θ_2 in equations (7-2) and (7-3). It's an easy procedure to tap off the outputs of these four integrators, plug them into a Mux block, and write a simple function that computes the error. Figure 7-7 shows the result of the Scope block that monitors this computed error over the course of a 10-second simulation of the robot falling under its own weight.

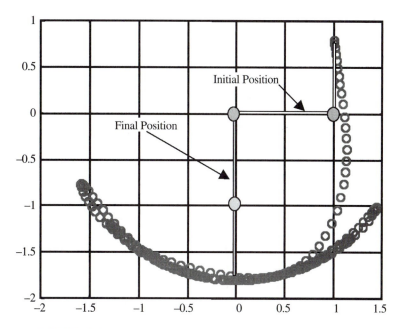

FIGURE 7-6 Results of simulation in which the robot is unpowered and the manipulator is allowed to fall under its own weight.

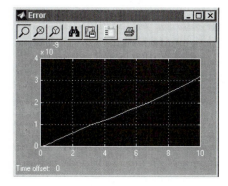

FIGURE 7-7 Scope block output of the norm of the error vector for the robot simulation.

Note that the error, while increasing steadily over time, starts at exactly zero—which indicates that the initial conditions were well-posed—and never achieves a value greater than 10E-8 meters over the course of 10 seconds of simulation time.

7.6 Robot Coordinate Control

Detailed design and implementation of robot control systems lie outside the scope of this book. However, the interested student can build upon this simulation and implement simple PD-loops (proportional plus derivative control loops) controlling the position of the individual joint motors (which is a typical industrial implementation). Alternately, more sophisticated schemes using a "computed torque" or inverse Jacobian could be investigated. In any event, the flexibility of the Simulink environment makes it possible to rapidly prototype various approaches to control and investigate their effectiveness.

It is left to the student to implement the portions of the simulation required to control this manipulator.

7.7 Summary

In this chapter, a fully dynamic model of a two-link robot is derived and implemented in the Simulink environment. A simulation of this type is very difficult to implement in a more general environment, because the closed-form equations of motion are very complicated and difficult to implement correctly. In addition, a simulation of this type is extremely useful in understanding the complex dynamics of serial chain manipulators and the difficulties encountered in controlling these systems.

CHAPTER 7 PROBLEMS

1. Use the dynamic simulation presented in this chapter to explore the behavior of this rather simple mechanical system. In particular, show that the inertia matrix contains off-diagonal terms. The main implication of off-diagonal terms in the inertia matrix is that force input in one joint will cause motion in the other.

2. Expand the dynamic simulation of the two-link robot presented in this chapter to include simple PD (proportional-derivative) control loops on each joint motor. The nature of PD control is that the torque generated by the motor is the sum of one signal that is proportional to the error between the desired and actual joint angles and another signal that is proportional to the derivative of that error. The constants of proportionality, known as gains, have a very large impact on the behavior of the system and must be tuned to achieve adequate performance. Spend some time tuning these control loops so that the robot can move swiftly from one point to another, but with no oscillation.

3. Equation (7-6) shows the Jacobian matrix that relates the payload motion to the joint motion. Modify the simulation and write an appropriate m-file that will use this relationship to control the robot by explicitly specifying the desired payload velocity.

CHAPTER 8

SIMULATING MECHANISMS THAT CHANGE

The previous chapters have dealt with the analysis and simulation of single degree-of-freedom mechanisms in which the nature of the kinematic constraints remains constant in time. In other words, the link lengths remain constant and the joints are always connected. Many interesting systems, such as legged robots or mechanisms with clearance in the joints, must make use of a more detailed analysis so that we might assemble appropriate simulations. In this chapter, we will look at a mechanism that changes over time and the methods required to analyze it.

8.1 The Geneva Mechanism

Figure 8-1 is a drawing of the Geneva mechanism. This well-known mechanism is used to impart intermittent motion with a continuously rotating driving source. The wheel on the right is driven at constant speed, and the pin near the periphery of the wheel engages one of the slots on the left wheel. While the pin and slot are engaged, they act like a prismatic (sliding) joint, and the slotted wheel moves 90°, at which point the pin leaves the slot and the slotted wheel stops and awaits the next rotation of the driving wheel.

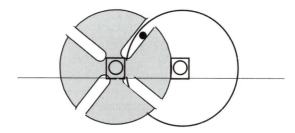

FIGURE 8-1 The Geneva wheel mechanism.

The key to simulating this system is to realize that while the pin and slot are engaged, the mechanism is kinematically equivalent to an inverted slider-crank mechanism. When they are not engaged, they are simply two wheels, one moving at a constant speed, the other stationary. A successful simulation of this mechanism, therefore, will have two sets of equations—one describing the engaged pair, one describing them when not engaged—and a set of conditions to decide which set of equations are in force at any given time.

Figure 8-2 shows an inverted slider crank in a configuration analogous to the Geneva wheel in Figure 8-1. The pinned wheel, shown on the right in Figure 8-1, is link 2 and the slotted wheel is link 3. The slider represents the pin in the slot. From the table in Chapter 2, the equations are easily derived.

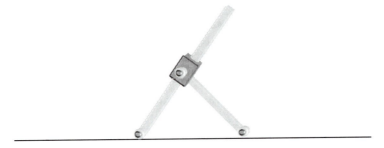

FIGURE 8-2 Inverted slider-crank mechanism analogous to the Geneva wheel while the pin and slot are engaged.

$$\begin{bmatrix} \cos\theta_3 & -r_3\sin\theta_3 \\ \sin\theta_3 & r_3\cos\theta_3 \end{bmatrix}\begin{bmatrix} \ddot{r}_3 \\ \alpha_3 \end{bmatrix} = \begin{bmatrix} -\alpha_2 r_2\sin\theta_2 - r_2\omega_2^2\cos\theta_2 + 2\dot{r}_3\omega_3\sin\theta_3 + r_3\omega_3^2\cos\theta_3 \\ \alpha_2 r_2\cos\theta_2 - r_2\omega_2^2\sin\theta_2 - 2\dot{r}_3\omega_3\cos\theta_3 + r_3\omega_3^2\sin\theta_3 \end{bmatrix}$$

This set of equations can quite easily be embedded in a kinematic simulation as described in Chapter 4. Figure 8-3 shows a likely configuration.

From the point of the beginning of engagement, the driving wheel moves through 90° of rotation, at which point the pin leaves the slot and a different set of equations hold true until the wheel completes a rotation and the pin engages in the next slot. The equations that govern the kinematics for this period of motion are simply:

$$\alpha_3 = 0$$
$$\ddot{r}_3 = 0$$

The function embedded in the simulation in Figure 8-3 can easily be modified to check the relative positions of the mechanism and use the appropriate equations to compute the accelerations. For reasons that will soon be clear, we find it useful to write a new function that monitors θ_2 and sets an output value to be 1 or 0 depending upon whether or not the pin is engaged. The function, called detect() is shown below:

detect.m

```
function trigg=detect(u)
%
%  Function to detect engagement of the Geneva Wheel
%  It assumes a symmetric arrangement of two wheels of the
%  same size and a 90 degree intermittent motion.
%
```

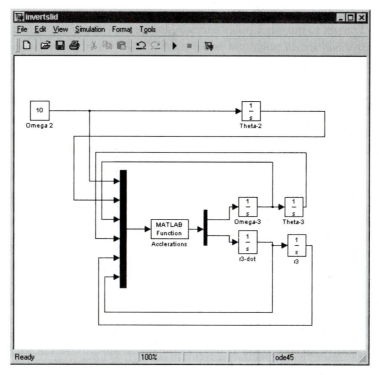

FIGURE 8-3 Simulink diagram showing the blocks needed for a kinematic simulation of an inverted slider crank (file: `invertslid.mdl`).

```
%  The function signals engagement (output goes to 1.0)
%  if the driving wheel (right-hand wheel) is between 135
%  and 225 degrees
%
ang=mod(u(1),2*pi);
if (ang>(135*pi/180) & ang<(225*pi/180))
   trigg = 1.0;
else
   trigg = 0.0;
end
```

This function is used to detect whether or not the pin is engaged and the output of this function is an input to the function that computes the acceleration.

Now we are almost finished with the kinematic simulation of the Geneva wheel. The only thing left is to consider the issue of initial conditions and the meanings of the integrator outputs in this case. For the integrators labeled θ_2 and ω_3, the initial conditions are relatively straightforward. We can assume $\theta_2 = 0$ at the beginning of our simulation and also $\omega_3 = 0$, since the pin is not engaged when $\theta_2 = 0$. The other three integrators, however, present special difficulties and will be discussed in greater detail.

The two integrators associated with the vector $\mathbf{R}_3$ must be considered very carefully. For the time when the pin is engaged, they represent the velocity and displacement of the pin relative to the slot. At other times, they should be zero. When the pin begins engagement, the signal generated by the detect function switches from 0 to 1 and r_3 should be at its maximum value (and equal to r_2 for most Geneva wheels) while the relative velocity will be the same as the absolute velocity of the pin, since it enters the slot in the tangent direction. Each time the pin enters a slot, these two integrators must be reset to these values. While this seems to be an impossible problem, the Simulink integrators have some special features that handle such situations quite well. The integrators can be configured so that they can be reset by a transition in a signal (like the change from 0 to 1 in the detect signal). Moreover, the user can choose between resetting the integrator to the value specified in the dialog box or to a value that is present on another input line. Therefore, we configure the integrators labeled "r3" and "r3-dot" to reset on a rising signal. For the displacement integrator we choose to use the value specified in the dialog box, the "internal" value of the initial condition. However, for the velocity integrator, the initial condition depends on the velocity of the driving wheel at the moment of engagement. Therefore, it is decided that the external property is chosen and the initial condition is computed as a function of ω_2.

Finally, we consider the problem of θ_3. For the Geneva mechanism, the slotted wheel will advance intermittently, 90° at a time. From an initial condition of, say, $\pi/4$ radians (as shown in Figure 8-1), it will advance to $-\pi/4$ for the first rotation of the pinned wheel. (Note that if the driving wheel moves counterclockwise, then the slotted wheel moves clockwise and the angles reflect negative motion.) The next rotation will take the slotted wheel from $-\pi/4$ to $-3\pi/4$, and so on. In contrast, the analogy of the inverted slider crank rests on the notion that the driven link moves from $\pi/4$ to $-\pi/4$ for one cycle of engagement. If the angle θ_3, as computed in our simulation, is fed directly back to the function that computes the acceleration (as shown in Figure 8-3), then it will be in error three out of every four times the pin is engaged. To solve this problem, we introduce one more integrator that will be reset, with the rising of the detect signal, to $\pi/4$. This value, which is labeled θ_3-hat in the simulation, will be used by the function to compute the accelerations, thus making the inverted slider-crank analogy correct for all times.

The final Simulink diagram is shown in Figure 8-4.

The function to compute the accelerations is listed below:

genacc.m

```
function out=genacc(u)
%
%  function to compute accelerations of a geneva wheel
%
%  u(1) = omega-2
%  u(2) = theta-2
%  u(3) = omega-3
%  u(4) = theta-3-hat (for equiv inverted slider crank)
%  u(5) = r3
%  u(6) = r3-dot
%  u(7) = engagement trigger
```

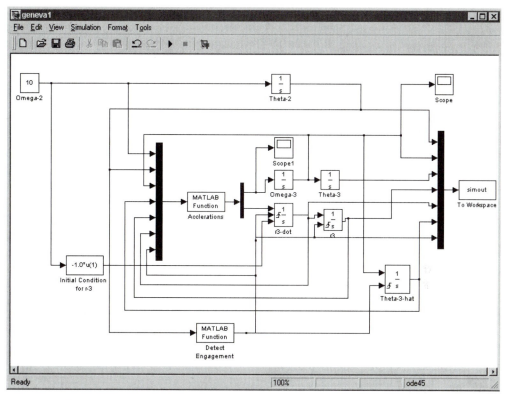

FIGURE 8-4 Completed Simulink diagram for the Geneva mechanism in Figure 8-1. In particular, note the integrators, which are reset by the function that detects whether or not the pin is engaged in a slot of the mechanism (file: geneva1.mdl).

```
%
r2=1.0;
%
if u(7) == 0
   out(1)=0.0;
   out(2)=0.0;
else
   S2=sin(u(2));
   C2=cos(u(2));
   S3=sin(u(4));
   C3=cos(u(4));
   a=[u(5)*C3 S3   ;-u(5)*S3   C3];
   b=[-r2*u(1)^2*S2-2*u(6)*u(3)*C3+u(5)*u(3)^2*S3;
      -r2*u(1)^2*C2+2*u(6)*u(3)*S3+u(5)*u(3)^2*C3];
   out=inv(a)*b;
end
```

Finally, Figures 8-5 and 8-6 show the motion of the slotted wheel, given a speed of 10 rad/s of the driving wheel.

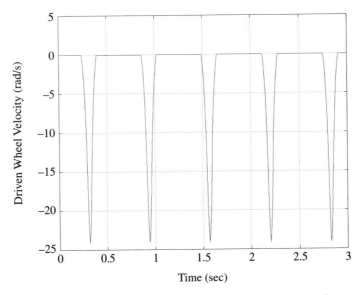

FIGURE 8-5   Velocity of driven wheel of the Geneva mechanism as computed by the kinematic simulation.

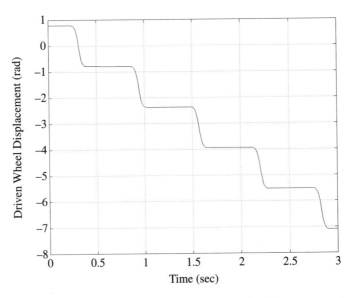

FIGURE 8-6   Angular displacement of the driven wheel of the Geneva mechanism as computed by the kinematic simulation.

8.2 Summary

In this chapter, we make use of some of the advanced features of MATLAB and Simulink to perform kinematic simulations of systems that change topology over time. The Geneva mechanism is alternately a system of two decoupled wheels and a kinematic analog to an inverted slider crank. This chapter presented one approach to simulating such a mechanism.

CHAPTER 9

THE TREBUCHET

9.1 Introduction

The trebuchet (pronounced *treb-YOO-shay*) is an ancient device that dates back to the Middle Ages and succeeded the better-known catapult. Like the catapult, the trebuchet was used to hurl large masses (such as stones) through great distances and to great heights. The main application was that of a siege engine. An army laying siege to a walled city could lob missiles over the walls, thus making life difficult to those inside attempting to wait out their besiegers.

Figure 9-1 shows a sketch of the trebuchet, which can be built to a variety of scales from a tabletop size that can throw pencil erasers to a monster capable of throwing automobiles.[1]

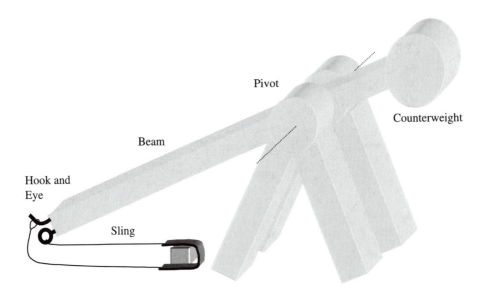

FIGURE 9-1   The trebuchet.

[1]Chevedden, P. E., et al. "The Trebuchet," *Scientific American*, Vol. 273, No. 1 (July 1995), pp. 66–71.

FIGURE 9-2 Sequence showing the trebuchet in operation. Starting at the top, the trebuchet is cocked and ready to fire. Once it is released, the counterweight causes the beam to rotate, which in turn whips the sling with its payload around, eventually releasing it to the right.

Figure 9-1 illustrates the trebuchet in the cocked position, ready for firing. Figure 9-2 is a sequence of sketches showing the trebuchet in operation. Once it is released, the counterweight is drawn down by gravity, causing the beam to rotate in a clockwise fashion. The payload is initially dragged to the left along the ground until the hook and eye lift the payload off the ground with the sling. By this time, the payload has gained a large amount of momentum in the leftward direction, which is converted to rotational momentum by the sling. As the motion proceeds, the payload gains velocity and eventually overtakes the beam, causing the rope loop to slip off the hook at the end of the beam, thus opening the sling and freeing the payload, which by this time is moving to the right and upward at a very high velocity. Armies in the Middle Ages were capable of launching very large payloads (thousands of pounds) over great distances with this ingenious device.

The trebuchet is a particularly good candidate for dynamic simulation because it is not easily analyzed by traditional kinematic methods. It is by its very nature a dynamic mechanism with no constant velocity members, no input other than the forces of gravity, and no easy way to predict the release of the sling. However, this device is highly sensitive to the geometric and inertial parameters of its various components. Short of actually building and testing the device, the only way of analyzing the trebuchet is through a fully dynamic analysis. The methods described in this book can be easily applied to a computer-based simulation using MATLAB and Simulink, as will be shown presently.

9.2 The Vector Loop

The vector loop for the trebuchet is quite simple, and the constraint equations are easily derived. Figure 9-3 shows the trebuchet in mid-motion and the pertinent displacement vectors are shown.

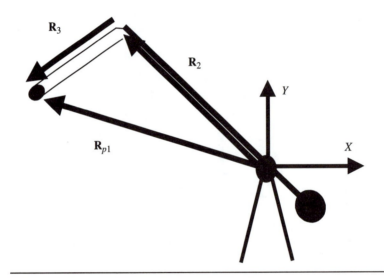

FIGURE 9-3 The trebuchet, showing the displacement vectors to describe its motion.

Note that the origin of the coordinate system is taken to be the pivot point. The vector loop equation is, therefore:

$$\mathbf{R}_2 + \mathbf{R}_3 = \mathbf{R}_4 \tag{9-1}$$

Note that, in writing this equation, we are assuming that the sling stays in tension and the rope loop remains on the hook. Later we will consider the case in which the rope slips off and the payload begins its flight. Also note that the center of gravity of the beam (link 2) will not, in general, lay on the line between the origin of the link's displacement vector and the tip of the vector. This will be easily handled by making r_{2c} the distance from the base of the vector to the center of mass, a negative number. The x- and y-components of the vector loop equation are:

$$r_2 \cos \theta_2 + r_3 \cos \theta_3 = x_{\text{payload}} \tag{9-2}$$

$$r_2 \sin \theta_2 + r_3 \sin \theta_3 = y_{\text{payload}} \tag{9-3}$$

Taking the derivative with respect to time twice yields the acceleration equations.

$$\ddot{x}_{\text{payload}} + r_2 \alpha_2 \sin \theta_2 + r_3 \alpha_3 \sin \theta_3 = -r_2 \omega_2^2 \cos \theta_2 - r_3 \omega_3^2 \cos \theta_3 \tag{9-4}$$

$$\ddot{y}_{\text{payload}} - r_2 \alpha_2 \cos \theta_2 - r_3 \alpha_3 \cos \theta_3 = -r_2 \omega_2^2 \sin \theta_2 - r_3 \omega_3^2 \sin \theta_3 \tag{9-5}$$

Since the center of gravity (cg) of the beam is not coincident with the pivot (about which we will be taking the torque equation), we will also need to relate the acceleration of the cg with the angular acceleration and velocity of the beam. In this case (as seen previously), the vector equation is somewhat trivial.

$$r_{2c} \cos \theta_2 = x_{2c} \tag{9-6}$$

$$r_{2c} \sin \theta_2 = y_{2c} \tag{9-7}$$

Taking the two derivatives with respect to time, we can derive the acceleration equations:

$$\ddot{x}_{2c} + r_{2c} \alpha_2 \sin \theta_2 = -r_{2c} \omega_2^2 \cos \theta_2 \tag{9-8}$$

$$\ddot{y}_{2c} - r_{2c} \alpha_2 \cos \theta_2 = -r_{2c} \omega_2^2 \sin \theta_2 \tag{9-9}$$

9.3 The Equations of Motion

Figure 9-4 illustrates the free-body diagram of the beam, showing all forces acting on it. Note that we drew it with a positive θ_2 in the first quadrant, which makes it easier to correctly account for the sign of the forces.

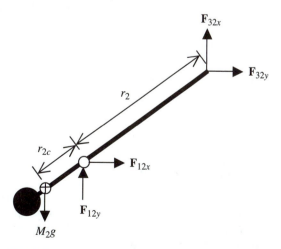

FIGURE 9-4 Free-body diagram of the trebuchet beam.

Now we write the three equations of motion for the beam.

$$F_{12x} + F_{32x} = M_2 \ddot{x}_{2c} \tag{9-10}$$

$$F_{12y} + F_{32y} - M_2 g = M_2 \ddot{y}_{2c} \tag{9-11}$$

$$-F_{32x} r_2 \sin\theta_2 + F_{32y} r_2 \cos\theta_2 - M_2 g r_{2c} \cos\theta_2 = I_2 \alpha_2 \tag{9-12}$$

Similarly, Figure 9-5 shows the free-body diagram for the sling. Note that the sling is a rope or cable, not a rigid link. Therefore, it is incapable of transmitting a torque. Also, since the mass of the rope or cable will, in general, be much smaller than the mass of the beam and payload, we will neglect the inertial effects of the sling and assume the payload is a point mass located at the end of the sling.

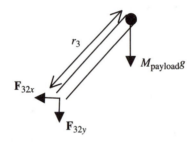

FIGURE 9-5 Free-body diagram of the sling.
For the purposes of this analysis, we assume
that the sling is in tension.

The following three equations are easily written:

$$-F_{32x} = M_{\text{payload}} \ddot{x}_{\text{payload}} \tag{9-13}$$

$$-F_{32y} - M_{\text{payload}} \, g = M_{\text{payload}} \ddot{y}_{\text{payload}} \tag{9-14}$$

$$-F_{32x} r_3 \sin\theta_3 + F_{32y} r_3 \cos\theta_3 = 0 \tag{9-15}$$

This last equation, which reflects the fact that the sling cannot react torques, can be manipulated to the following, slightly simpler form.

$$F_{32y} - F_{32x} \tan\theta_3 = 0 \tag{9-16}$$

9.4 The Matrix Equation

In the previous two sections, we derived 10 independent equations with 10 unique unknowns (equations 9-4 through 9-15). The next step is to place the equations in matrix form so that we can enlist MATLAB's matrix routines to solve them within the simulation.

$$
\begin{bmatrix}
1 & 0 & r_2 S_2 & r_3 S_3 & 0 & 0 & 0 & 0 & 0 & 0 \\
0 & 1 & -r_2 C_2 & -r_3 C_3 & 0 & 0 & 0 & 0 & 0 & 0 \\
0 & 0 & r_{2c} S_2 & 0 & 1 & 0 & 0 & 0 & 0 & 0 \\
0 & 0 & -r_{2c} C_2 & 0 & 0 & 1 & 0 & 0 & 0 & 0 \\
0 & 0 & 0 & 0 & M_2 & 0 & -1 & 0 & -1 & 0 \\
0 & 0 & 0 & 0 & 0 & M_2 & 0 & -1 & 0 & -1 \\
0 & 0 & I_2 & 0 & 0 & 0 & 0 & 0 & r_2 S_2 & -r_2 C_2 \\
M_{pl} & 0 & 0 & 0 & 0 & 0 & 0 & 1 & 0 \\
0 & M_{pl} & 0 & 0 & 0 & 0 & 0 & 0 & 0 & 1 \\
0 & 0 & 0 & 0 & 0 & 0 & 0 & 0 & -S_3 & C_3
\end{bmatrix}
\begin{bmatrix}
\ddot{x}_{payload} \\
\ddot{y}_{payload} \\
\alpha_2 \\
\alpha_3 \\
\ddot{x}_{2c} \\
\ddot{y}_{2c} \\
F_{12x} \\
F_{12y} \\
F_{32x} \\
F_{32y}
\end{bmatrix}
=
\begin{bmatrix}
-r_2 \omega_2^2 C_2 - r_3 \omega_3^2 C_3 \\
-r_2 \omega_2^2 S_2 - r_3 \omega_3^2 S_3 \\
-r_{2c} \omega_2^2 C_2 \\
-r_{2c} \omega_2^2 S_2 \\
0 \\
-M_2 g \\
-M_2 g r_{2c} C_2 \\
0 \\
-M_{pl} g \\
0
\end{bmatrix}
$$

$$(9\text{-}17)$$

9.5 The Dynamic Simulation

The equations in the last section describe the dynamics of the trebuchet for the period of time that it is imparting momentum to the payload. It would be quite instructive as well to follow the payload after the sling is released and while the payload is in ballistic trajectory.

In this situation, we will ignore the sling and compute accelerations only for the payload and the beam. The equations are shown below:

$$\ddot{x}_{\text{payload}} = 0 \tag{9-18}$$

$$\ddot{y}_{\text{payload}} = -g \tag{9-19}$$

$$\alpha_2 = -M_2 / I_2 \, g \, r_{2c} C_2 \tag{9-20}$$

Detecting the exact moment at which the sling is released can be highly problematic. We choose to bypass this issue by assuming that the hook on the end of the beam has been adjusted, by trial and error, to release the sling when the payload has a velocity such that its angle of flight is 45° elevated from the horizontal, thus giving the payload its maximum distance.

Figure 9-6 shows the Simulink simulation of the trebuchet. There are two function files called by this simulation. The first, `trebsys.m`, is the function used to solve the simultaneous constraint matrix shown in equation (9-17). The second, `isreleased.m`, is a function used to set the release of the payload.

Note that the simulation is rather crowded and it is becoming difficult to follow the lines. At this point it would make sense to group functionally related blocks and hide them in a subsystem. That exercise is left to the student.

There are two elements in the Simulink diagram that are new to this book. First, the `Stop` block is inserted to automatically stop the simulation when the payload returns to earth. Second, a `Display` block gives us a readout of the distance that the projectile travels down range. The `X-Y Graph` block will give us a graph of the projectile's motion as well.

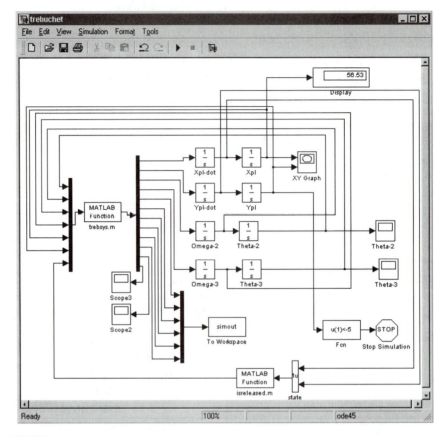

FIGURE 9-6 Simulink implementation of the trebuchet simulation (file: `trebuchet.mdl`).

The m-file that computes the simultaneous constraints is listed here:

trebsys.m

```
function [out]=trebsys(u)
%
%  u(1) = Theta-2
%  u(2) = Omega-2
%  u(3) = Theta-3
%  u(4) = Omega-3
%  u(5) = Xpayload
%  u(6) = Ypayload
%  u(7) = release switch; 0 implies normal; 1 implies release
%
r2=3.0;
r3=1;
r2c=-0.5;
```

```
m2=500.0;
I2=100.0;
mpl=5.;
g=9.81;
%
S2 = sin(u(1));
C2 = cos(u(1));
S3 = sin(u(3));
C3 = cos(u(3));
%
if u(7)==0
a=zeros(10,10);
%
a(1,1)=1;a(1,3)=r2*S2;a(1,4)=r3*S3;
a(2,2)=1;a(2,3)=-r2*C2;a(2,4)=-r3*C3;
a(3,3)=r2c*S2;a(3,5)=1;
a(4,3)=-r2c*C2;a(4,6)=1;
a(5,5)=m2;a(5,7)=-1;a(5,9)=-1;
a(6,6)=m2;a(6,8)=-1;a(6,10)=-1;
a(7,3)=I2;a(7,9)=r2*S2;a(7,10)=-r2*C2;
a(8,1)=mpl;a(8,9)=1;
a(9,2)=mpl;a(9,10)=1;
a(10,9)=-S3;a(10,10)=C3;
b(1)=-r2*u(2)^2*C2-r3*u(4)^2*C3;
b(2)=-r2*u(2)^2*S2-r3*u(4)^2*S3;
b(3)=-r2c*u(2)^2*C2;
b(4)=-r2c*u(2)^2*S2;
b(5)=0;
b(6)=-m2*g;
b(7)=-m2*g*r2c*C2;
b(8)=0;
b(9)=-mpl*g;
b(10)=0;
%
%
out=inv(a)*b';
%
errx=r2*C2+r3*C3-u(5);
erry=r2*S2+r3*S3-u(6);
out(11)=errx;
out(12)=erry;
else  % released
   out(1)=0.0;
   out(2)=-g;
   out(3)=-m2*g*r2c*C2/I2;
   out(4)=0.0;
   out(5:12)=[0 0 0 0 0 0 0 0];
end
```

Finally, the file to detect the release of the payload is listed here:

isreleased.m

```
function sw=isreleased(u)
%
% u(1) = Vx-payload
% u(2) = Vy-payload
%
ang=atan2(u(1),u(2));
if ang > pi/4
    sw=1;% released
else
    sw=0;
end
```

9.6 Simulation Results

Table 9-1 shows some typical parameters for a large-scale trebuchet.

TABLE 9-1   Inertial and Geometric Parameters for a Trebuchet

Parameter	Value	Parameter	Value
M_2	500 kg	r_2	3.0 m
M_{pl}	100 kg	r_{c2}	–0.5 m
I_2	10 kg m^2	r_3	1.0 m

At the cocked position, the initial conditions are predetermined and are summarized in the Table 9-2.

TABLE 9-2   Consistent Initial Conditions for the Trebuchet Simulation

Integrator	Value	Integrator	Value
xpl-dot	0	X_{pl}	1.12132 m
ypl-dot	0	Y_{pl}	–2.12132 m
ω_2	0	θ_2	255°
ω_3	0	θ_3	0

Figure 9-7 shows the trajectory of a payload of 10 kg.

Similarly, for a 1 kg payload, the simulation predicts the trajectory shown in Figure 9-8 and computes that the payload would land over 128 m down range.

9.7 Summary

The trebuchet is a deceptively complex device. It has only one revolute joint and yet gives rise to very complex dynamic behavior. None of the motion-based methods normally covered in undergraduate kinematics classes could be used to analyze this ma-

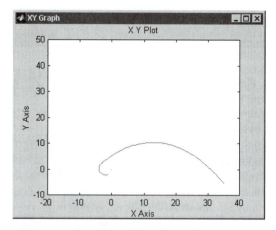

FIGURE 9-7 Payload trajectory for a payload of 10 kg.
Note that the trebuchet moved the projectile over 35 m down range.

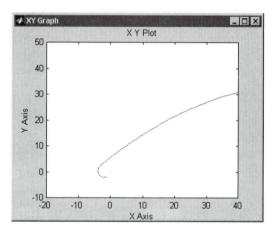

FIGURE 9-8 Projectile trajectory for a 1 kg payload.
This projectile landed over 128 m down range.

chine. There are no parts operating at (or near) constant speed. The machine is inherently dynamic and must be analyzed that way.

As was the case with all mechanisms modeled in this book, the techniques presented here are merely starting points. The simulations can (and should) be expanded and built upon to be used as design tools or sophisticated analysis models.

This book attempted to provide the tools and guidance to allow the student and practitioner to make use of MATLAB or similar simulation software to quickly implement fully dynamic simulations of constrained systems without the use of sophisticated energy-based methods that are typically restricted to graduate-level understanding of dynamics. The background required for this approach is simple engineering vector mechanics and the fundamentals of matrix operations. It is hoped that the reader will find this method useful for a wide range of engineering applications.

CHAPTER 9 PROBLEMS

1. The engineers who worked with trebuchets soon learned that modifications could be made to the basic design that would improve performance. One such modification is the inclusion of a hinged bucket in the place of a fixed counterweight. This allowed the rotational inertia of the beam to be kept low, allowing higher throwing velocities. Modify the simulation to allow for this design improvement, and verify that it does indeed improve performance.

2. Another important modification is to mount the entire device on wheels, allowing it to roll along the ground in the same direction that it throws. Not only did this improve the mobility of the device, but it could also improve performance. Modify the dynamic simulation to allow the base to move horizontally on wheels. Experiment with the mass of the base relative to the counterweight to find the best combination for performance.

3. Create a simulation of the trebuchet, implementing both design improvements. Use the resulting simulation as a design tool to find the best set of parameters for delivering a 250 lb payload.

SIMULINK TUTORIAL

This tutorial is intended to bring a student up to speed on the basics of Simulink in a short period of time. Simulink is a versatile and powerful analysis tool with a rich feature set and powerful numerical algorithms. The online documentation (available through the helpdesk command in MATLAB) provides a far more complete description, and users are encouraged to familiarize themselves with this feature.

A.1 Starting Simulink

The MATLAB window upon startup looks like the window shown in Figure A-1.

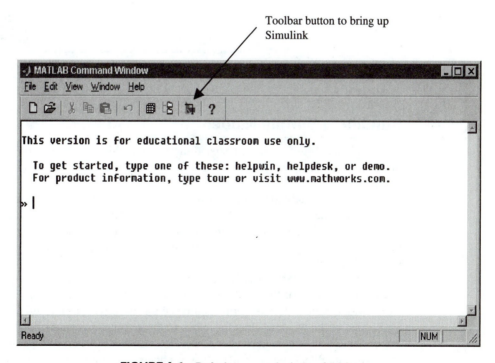

FIGURE A-1 Default command window for MATLAB.

Note the Simulink button on the toolbar in the default widow. If this button is not visible, then Simulink is not installed on your machine. Contact your system administrator for help. When you click on the Simulink button, you will bring up the window shown in Figure A-2.

Toolbar button to open
new Simulink window

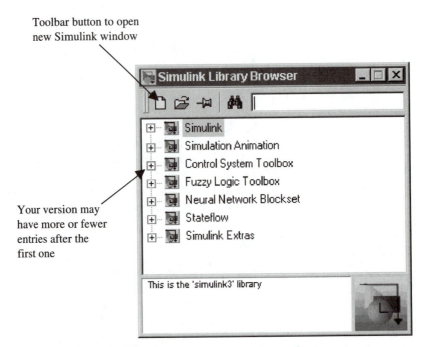

Your version may
have more or fewer
entries after the
first one

FIGURE A-2   Simulink Browser Library window.

A.2   Building a Simple Model

Now, press the "New" icon, and expand the Simulink block library by pressing on the "+" icon in the Simulink window. Your screen should show the windows as indicated in Figure A-3.

We are now ready to prepare our first simulation. At the core of computer simulation lies numerical integration. Therefore, we will begin with an example of simply integrating a signal. The simulation will contain an input (or signal "Source," in Simulink parlance), two integrators, and a window in which we can watch the signals change with time (a "Sink").

Now, go to the Simulink Library Browser window (the window that appeared when you started you Simulink session). Expand the "Source" library and find the `Step` block. Click on it and you should see a brief description of this block in the lower pane of the window. Drag this block to the empty window by clicking and holding on the `Step` block, and move it to the empty window while holding down the mouse button. When you get to the window, release the mouse button, and a copy of the `Step` block appears in your window (Figure A-4).

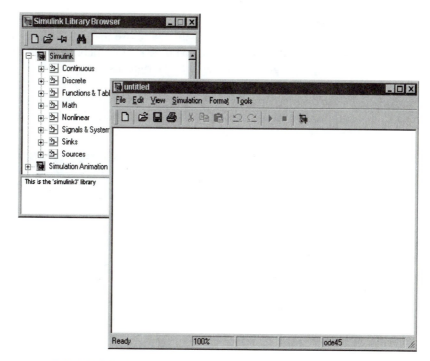

FIGURE A-3   Untitled new model window for new Simulink model.

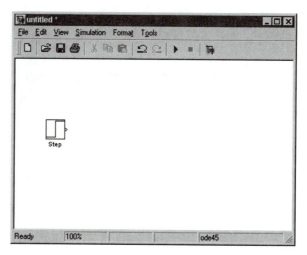

FIGURE A-4   Begin building the Simulink model by dragging the Step block to the new model window.

Similarly, find the `Integrator` block under the "Continuous" library and move an integrator into the window. We will require two integrators in this simulation, but drag only one from the library. Once you have the integrator next to the step function, right-click on the integrator and drag it to the right a short distance within the same window and release. You should now see two integrators. This is a very useful feature of Simulink; you can create a copy of any block by right-clicking and dragging to a new location.

From the "Sinks" library, find the `Scope` block and drag it onto the window. Your simulation should now look like Figure A-5.

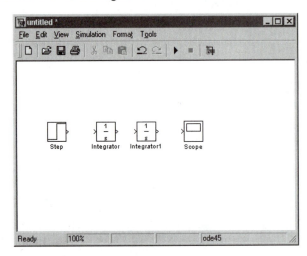

FIGURE A-5　Continue building the model by adding
two `Integrator` blocks and one `Scope` block.

The block diagram is completed by connecting the blocks. Use the mouse to drag connecting lines from one block to the next (in much the same way you would draw a line in a drawing program). Finally, the simulation should look like Figure A-6.

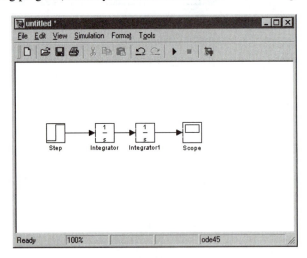

FIGURE A-6　Connect the blocks by "drawing" connecting lines.

Now would be a good time to save your work. File management is done in the manner consistent with all Windows® programs (or Macintosh, Unix, or Linux, if that is the version you are using). Click on the disk icon, or pull down the File menu and select Save As to name the file. Note that after you've done so, the name of the file appears on the window header. The default extension for Simulink model file names is .mdl.

A.3 Running the Simulation

There are more details to be examined, but at this point we've defined the minimum required for a simulation. Double-click on the Scope icon and press the run button on the Simulink window. The run button looks like the play button on a CD player. After a few moments, the end of the simulation is signaled by a muted beep, and the scope window should look like the one shown in Figure A-7.

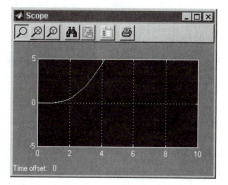

FIGURE A-7 Scope block output for first simulation run.

Several things may be observed from this figure. First, by default, Simulink runs a simulation from 0 to 10 seconds. Second, the output of the second integrator appears to be a quadratically increasing signal, but this behavior does not begin until after 1 second of simulation time has passed. Finally, the default scale on the scope (± 5) is insufficient to capture this entire signal.

Now we will examine the various settings in Simulink and the Simulink blocks and see how they affect the simulation run.

A.4 Simulation Run-Time Parameters

In the simulation window (as shown in Figure A-6), go to the Simulation menu and select Parameters. The dialog box shown in Figure A-8 should appear.

Note that here is where you select the start time and stop time of the simulation. For the vast majority of physical simulations, the actual value you use for start time is not relevant. Only in those simulations in which there are explicit functions of time does it matter. Convention dictates that, in lieu of any information to the contrary, start the simulation at $t = 0.0$.

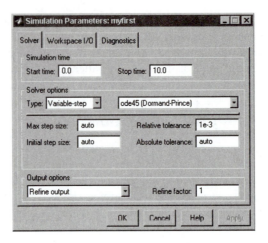

FIGURE A-8 Dialog box for the Simulation → Parameters menu selection.

This window is also used to adjust parameters that tell Simulink how to carry out the numerical integration that is the core of the simulation. By default, it uses a variable step-size 4th/5th order explicit integration scheme that is a refinement of the classic Runge-Kutta method. This method adjusts the integration time step in a manner that attempts to maintain an estimate of integration error below a certain tolerance. By default, the relative tolerance (error relative to the size of the integrator output) is set to 0.001 (one part in one thousand, or 0.1%). The absolute tolerance is automatically adjusted by the algorithm.

For the simulations discussed in this text, the relative tolerance should be tightened up considerably to, say, 1E-6 or 1E-7. Make that adjustment in the window now. Similarly, change the final time to 5.0 seconds, and close the window by clicking the OK button. If you run the simulation now, the scope will show only 5 seconds of simulated time and the signal will be plotted appropriately.

Now go back to your simulation model and open up the "step input" block by double-clicking on it. You should see a window similar to the one shown in Figure A-9.

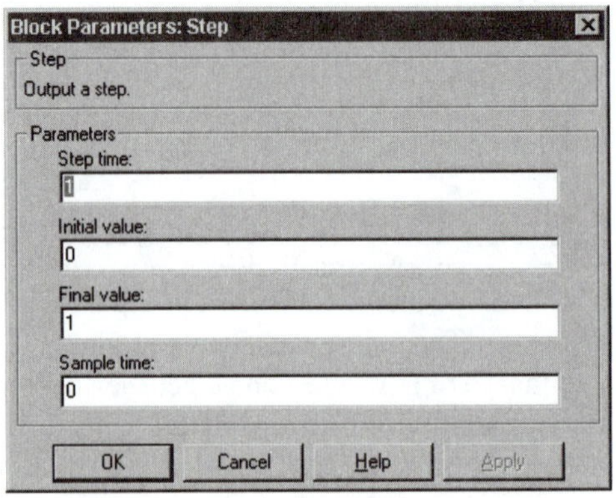

FIGURE A-9 Dialog box showing options for the Step block.

Here we see the reason why the simulation output began to change after 1 second had passed. The integrators were receiving zero input from 0 to 1 second. The `Step` block generates a signal that changes from its initial value to its final value at the "step time" entered in this window. The image on the block itself is a representation of how this signal would look on an oscilloscope, or a plot of the signal versus time. It appears as a stair step, hence the name.

Change the step time to 0.0 and close the dialog box.

A.5 Initial Conditions

Let's take a minute to reflect on this simple simulation. The `Step` block generates a signal that begins at zero, but as soon as the simulation begins, it becomes a signal equal to 1.0. The next two blocks integrate that signal with respect to time, and the result of this double integration is plotted in a window that simulates an oscilloscope. Mathematically, the process may be represented this way:

$$y(t) = \int_0^t \int_0^\tau u(\sigma)\, d\sigma\, d\tau$$

Note that, in keeping with the formalism of integral calculus, we make use of dummy variables for the integration process. Assuming that the function being integrated is constant at 1.0 for the time of the integration, we can carry out this integration analytically to find the following result:

$$y(t) = \frac{1}{2}t^2 + C_1 t + C_2$$

where C_1 and C_2 are the constants of integration we recall from integral calculus. In the context of numerical simulations, these constants of integration have important physical significance. They are the *initial conditions* of the values that the integrators are computing. In other words, they are the values of the signals at time $t = 0$. *All simulations require that the user provide correct initial conditions for each integration process (or integrator block) used in the simulation.* By default, the initial conditions of the integrators used in Simulink are 0.0, but they can be adjusted as we will soon see.

Now run the simulation and examine the results. You should now see that the output of the simulation begins changing at 0 seconds, due to the fact that we've modified the step input block. Take a moment to examine the scope output in Figure A-10, which should be the same as the plot you've just generated.

By examining this figure, we can see the effects of both initial conditions on this output. The fact that the value of the signal is zero at the initial time reflects the fact that the initial condition on the second integrator (righthand) is zero. The fact that the slope (or derivative) of the signal is zero at $t = 0$ reflects the fact that the initial condition of the first integrator is also zero. Double-click on the righthand integrator and examine the dialog box that appears, shown in Figure A-11.

For the time being, we are interested only in the initial conditions. Change the initial condition of this integrator to 1, click OK, and run the simulation again. The plot shown in Figure A-12 should appear.

Note that the signal is now offset by the amount equal to the initial condition and that the slope is also zero (since the IC of the first integrator was not changed). The

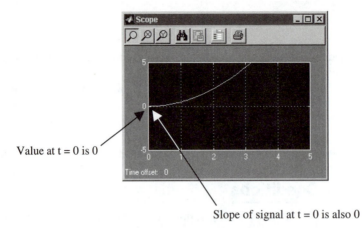

Value at t = 0 is 0

Slope of signal at t = 0 is also 0

FIGURE A-10   Scope block of simulation output for zero initial conditions and a step time of 0.0 seconds.

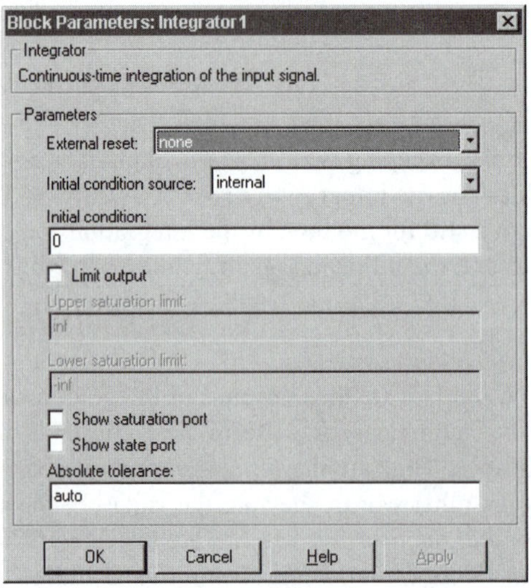

Block Parameters: Integrator1

Integrator

Continuous-time integration of the input signal.

Parameters

External reset: none

Initial condition source: internal

Initial condition:

0

☐ Limit output

Upper saturation limit:

inf

Lower saturation limit:

-inf

☐ Show saturation port

☐ Show state port

Absolute tolerance:

auto

| OK | Cancel | Help | Apply |

FIGURE A-11   Dialog box for the Integrator block.

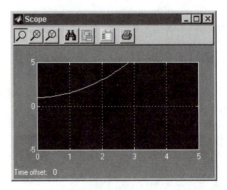

FIGURE A-12   Scope block for simulation run after the initial condition of the second integrator has been changed.

students are encouraged to experiment with both initial conditions to make sure that they understand these effects thoroughly.

A.6 Multiplexing Signals

Now, let's make some further modifications of the simulation to demonstrate a few more features of Simulink. Go back to the Simulink library browser, and under the "Signals & Systems" heading, find the Mux block and drag it to your simulation. Select the line that connects the last integrator to the scope block and delete it by pressing the delete button. Your simulation should now look like Figure A-13.

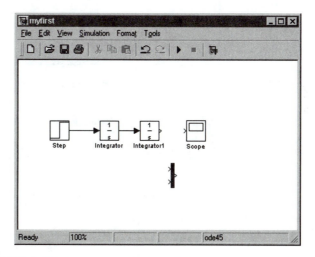

FIGURE A-13 Insert a Mux block and delete one of the connecting lines.

The Mux block is a "multiplexor" that takes multiple signals and merges them into a single line. It's a useful feature that we will use on many occasions. You can change the number of inputs by double-clicking and using the dialog box. Do that now and make it a three-input multiplexor. Rearrange the simulation so that the Mux now stands between the second integrator and the scope and connect the second integrator to the first Mux input and the output of the Mux to the Scope block. The simulation should now look like the model shown in Figure A-14.

To complete the modification, we will "tap" two of the existing signals and map them into the other two Mux inputs. To tap an existing signal line, bring the cursor to the line and right-click (command-click for Mac users), then draw the new line as you normally would. The window display indicates a connection between the original line and the tapped line (as opposed to crossing, unconnected lines) with a small filled box. The completed modification should look like Figure A-15.

Reset both initial conditions to zero and run the simulation and examine the results on the scope (Figure A-16).

Note that the three lines (which appear as different colors on your screen) indicate the three signals that were input to the Mux. The horizontal line is the third input, the

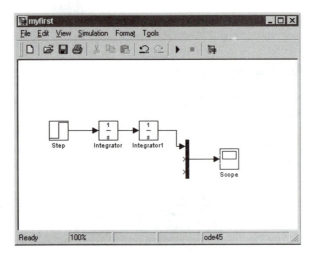

FIGURE A-14 Extend the Mux to accommodate three inputs and begin to connect it.

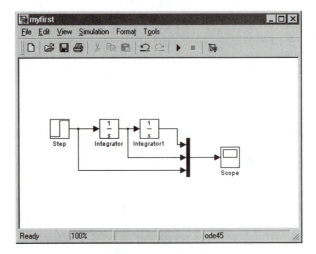

FIGURE A-15 The modified simulation allowing all signals to be viewed simultaneously.

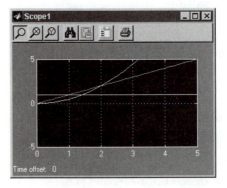

FIGURE A-16 Scope block with three simultaneous signals displayed.

step that changes to 1.0 at the beginning of the simulation and stays there throughout the simulation. The straight line that begins at zero and ends at a value of 5.0 when the simulation ends at 5 seconds is the second Mux input, which is the output of the first integration. A little reflection will show that this is an appropriate response for the integration of a constant value. Finally, the result of the second integrator is the same as we saw from our original simulation run.

At this point, the student is encouraged to adjust initial conditions, step time, and magnitude of the step input and study their effects until a solid intuitive understanding of these factors is gained.

A.7   Simulink and MATLAB: Returning Data to the Workspace

One of the most powerful features of Simulink is that it is part of the MATLAB environment and can exchange data seamlessly. To explore one possible connection between Simulink and the MATLAB workspace, replace the Scope block in our example with a To Workspace block (also found in the "Sinks" library). The new simulation is shown in Figure A-17.

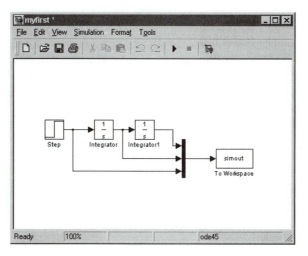

FIGURE A-17   Replace the Scope block with a To Workspace block to store simulation results directly to the MATLAB workspace (file: myfirst.mdl).

The To Workspace block will, in general, take a multisignal line as input and generate a matrix in the MATLAB workspace. Each row of the matrix represents the values of the input variables at a given time step. The variables appear in the matrix in the same order in which they are connected to the Mux. In other words, the output of the second integrator in Figure A-17 is connected to the topmost, or first, Mux input. Therefore, the first column in the workspace matrix will be the output of the rightmost integrator.

Now double-click on the To Workspace block and examine the resulting dialog box, shown in Figure A-18.

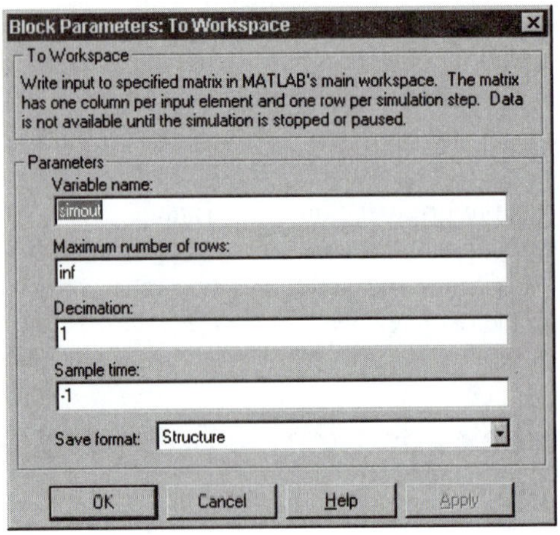

FIGURE A-18 Dialog box for the `To Workpace` block allowing users to set the name of the variable stored in the workspace.

The variable name refers to the name by which you can manipulate the resulting matrix in the MATLAB environment. Any qualified MATLAB matrix name will do. For now, we will keep the default name of `simout`. More important, however is the "Save format" near the bottom of the dialog box. The default (for R11 of MATLAB/Simulink) is "Structure." This format stores the data in a very compact and complete data structure. Unfortunately, it is also a bit ungainly to manipulate unless you are well versed in the concepts of data structures. For most of us, the other option in this box, "Matrix," is more suitable. Make this change in the box and run the simulation.

Since we've eliminated the scope, there will be no visual manifestation that the simulation is running, but a quiet beep will be heard when it is completed. Now go to the MATLAB command prompt window (the window that comes up when you first start MATLAB), and type in the command that lists the matrices stored in the workspace, `whos`.

```
» whos
  Name          Size          Bytes  Class
  simout        53x3           1272  double array
  tout          53x1            424  double array
Grand total is 212 elements using 1696 bytes
»
```

MATLAB shows two variables in the workspace (perhaps more if you have been doing other operations before you issue the command), beginning with the second matrix, `tout`. This matrix, which is 53 rows by 1 column, is generated by default by Simulink models and lists the times at each integration step. This is important because Simulink will, in general, utilize a variable time step algorithm and the actual times at which data are stored will not be known in advance.

The other variable, simout, is generated by the To Workspace block in our simulation. Note that the matrix is 53 rows by 3 columns, which is consistent with the fact that the Mux has three variables as input. To verify this fact, let's take a look at numbers stored in the matrix. The easiest way to do this is to type the name of the matrix at the prompt.

```
» simout
simout =
          0          0     1.0000
     0.0000     0.0000     1.0000
     0.0013     0.0505     1.0000
     0.0051     0.1010     1.0000
     0.0202     0.2010     1.0000
     0.0453     0.3010     1.0000
     0.0804     0.4010     1.0000
     0.1255     0.5010     1.0000
...
```

The first column, as previously mentioned, contains the output of the second integrator, a quadratically increasing variable. The second column is the signal between the integrators, the linearly increasing ramp function. The third column is the input variable.

A.8 Using the MATLAB plot command

The plot command in MATLAB is a powerful and flexible command. The complete details of this command are beyond the scope of this tutorial. However, a few pertinent examples will be helpful. The following command will generate a plot of the output of the second integrator versus time.

```
>> plot(tout,simout(:,1))
```

This is the easiest form of the plot command. A few comments regarding the syntax are in order. The ":" operator has unique properties in MATLAB. In this case, it selects the entire first column. The best way to understand this is to read the ":" operator in this case as "all." Therefore, the plot command shown above can be read as: "Plot all rows, first column of simout versus tout." The plot generated with this command is shown in Figure A-19. This figure was inserted into this document using the following command:

```
>> print -dmeta
```

This command copies the figure window to the clipboard, using the "windows metafile" format. Plots of multiple data sets can also be generated as shown in Figure A-20:

```
>>plot(tout,simout(:,1),tout,simout(:,2),tout,simout(:,3))
```

Finally, we will demonstrate the ability to affect the appearance of the plot using various characters to denote the data (as opposed to solid lines). See Figure A-21.

```
>>plot(tout,simout(:,1),'o',tout,simout(:,2),'+',tout,simout(:,3),'v')
```

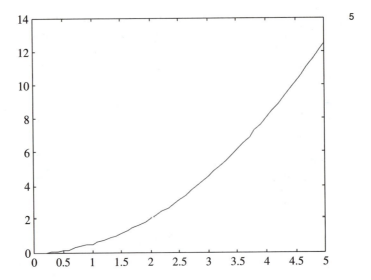

FIGURE A-19 Plot of integrator output versus time using the `plot` command in MATLAB.

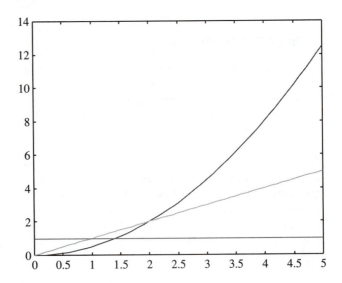

FIGURE A-20 Simultaneous plot of all three columns of the `simout` matrix versus time.

For more information on the `plot` command, type `help plot` at the command prompt.

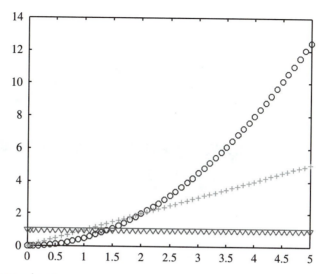

FIGURE A-21 Use of `plot` command with symbols for each data point as opposed to interpolated lines.

A.9 Using MATLAB Functions in Simulink

The methods demonstrated in this text make use of a powerful feature that combines the programming capabilities of the MATLAB environment with the highly visual approach used by Simulink to build simulations. The MATLAB Fcn block, found in the "Functions & Tables" library, can be used to incorporate any user-written MATLAB Function file into the simulation. This technique is perhaps best demonstrated by means of an example. It should be noted that the example used in this section—the response of a linear spring-mass-damper system—can be solved analytically and simulated by means other than a MATLAB function. We choose this example, however, because it is so well known in many fields of engineering.

A.9.1 An Aside: MATLAB Functions

We begin by reviewing the concept of a function as it pertains to MATLAB. In MATLAB, users can create text files of sequential MATLAB operations and call them from the command prompt. These files, called m-files due to the filename extension used (.m), may take on one of two forms. Script files are simply MATLAB commands that are replayed just as if they were entered at the command prompt. Any variable stored in the MATLAB workspace is available to script files. Functions, in contrast, operate in their own temporary workspace. Information is passed to functions as parameters, and information is returned to the workspace once the function is completed.

As an example, consider the factorial function, which is a function defined on nonnegative integers, defined as follows:

$$f(x) = x! = \begin{cases} 1 & \text{for } x = 0 \\ x*(x-1)*(x-2)*\cdots*1 & \text{for } x > 0 \end{cases}$$

We will now create a MATLAB function that computes factorials. From the default MATLAB command prompt, press the toolbar icon for new document or select New→ M-File from the File menu. This will open a new window that is the MATLAB editor and debugger; it should look like the window in Figure A-22.

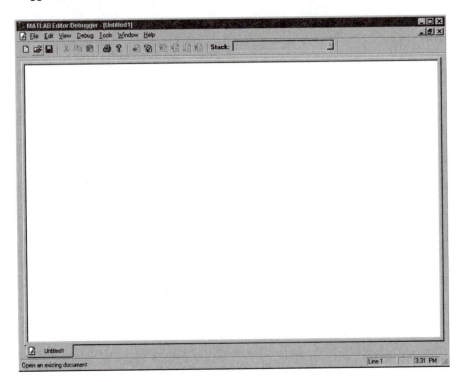

FIGURE A-22   The MATLAB editor/debugger window.

Now enter in the following lines of code. Note that the editor makes use of color to automatically highlight your code as you progress. Comments (anything to the right of a % sign) are shown in green, MATLAB keywords are in blue, and suspect errors are in red.

factorial.m
```
function  fact=factorial(u)
%
%  factorial function used to demonstrate the use of
%  function m-file in MATLAB
%
%
fact = 1;
if u>0
```

```
    for i=1:u
        fact=fact*i;
    end
end
```

The most important part of the function m-file, and the feature that distinguishes it from the script m-file, is the function declaration in the first line. The keyword `function` signifies that the m-file will act as a function. The assignment on the same line defines the name of the function (`factorial`), the name of the dummy variable that will be used to access parameters passed to the function (`u`), and the name of the variable that the function will use to return values to the calling statement (`fact`). Finally, it should be noted that this file is stored in an m-file of the same name as the function, namely, `factorial.m`.

As an example, let's use this function to compute the factorial of 5, using the definition of factorial:

$$5! = 5 * 4 * 3 * 2 * 1 = 120$$

From the MATLAB prompt, type:

```
» factorial(5)
ans =
   120
»
```

Note that the correct answer, 120, is stored in a variable called `ans` in the MATLAB workspace. The following form is also allowed:

```
» xf1=factorial(5)
xf1 =
   120
»
```

In this case, an assignment statement is used to call the function and store its returned value in the new variable, `xf1`.

Those who are familiar with computer programming principles can recognize the process here, but we will quickly summarize for the sake of clarity. The assignment statement, `xf1=factorial(5)`, directs MATLAB to take the value 5 and search for a function called `factorial`. If you've stored `factorial.m` in the default directory, MATLAB invokes it, takes the value 5, then in a memory workspace separate from the main MATLAB workspace, creates a new variable, called `u` (as designated in the function `declaration`), and stores the numerical value of 5 in it. The statements of the function are then executed. At the end of the function, whatever value is stored in the local variable `fact` (also designated in the function `declaration`) is returned to the workspace and stored in the variable `xf1`.

A very important yet subtle point is that the actual names of the variables used in the function file are entirely arbitrary. The "outside world" has no knowledge of how a particular function was programmed or what the variable names were. To demonstrate this point, the factorial function is reprogrammed using different names and is shown below:

```
function  res=factorial(udiff)
%
%  factorial function used to demonstrate the use of
%  function m-file in MATLAB
%
%
res = 1;
if udiff>0
   for i-1:udiff
      res=res*i;
   end
end
```

This behaves in precisely the same manner as the previous function file.

A.9.2 Calling Functions from Simulink

To demonstrate the combination of function m-files and Simulink we will construct a simple, if rather contrived, example similar to the previous example. Let's suppose that we wish to construct a simulation in which an arbitrary input is integrated twice, but that there is an algebraic function between the output of the first integrator (call it x_1) and the input of the second integrator (call it $\dot{x}_2$). This relationship is defined below:

$$\dot{x}_2 = 3x_1 - x_1^2$$

To implement this simulation, modify our existing simulation by inserting a MATLAB Fcn block (found in the Functions & Tables library) between the two integrators. Note that you break an existing signal link by selecting it and pressing the delete key.

Your simulation should now look something like the model shown in Figure A-23.

Since there is now one more signal to monitor (the output of the function), we'll increase the Mux to four inputs and connect both the input and output of the function block to the Mux.

Now we turn to the m-file editor and open a new document. The function we require is a single-input, single-output function that will implement the polynomial function shown above. Note that the actual choice of variable names is arbitrary, but MATLAB convention dictates that parameters are passed as the variable u.

One possible implementation of this function is shown below:

mydemo.m

```
function x2dot=mydemo(u)
%
%  function used to demonstrate the use
%  of functions in Simulink
%
x2dot=3*u - u^2;
```

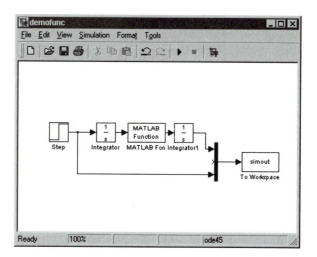

FIGURE A-23   Insert a MATLAB Fcn block between the two integrators
to demonstrate its use (file: demofunc.mdl).

Aside from comments, the function has only two lines, the function declaration and the
assignment statement carrying out the algebra.

Next, the Simulink model needs to be made aware of the function. Double-click on
the Fcn block to find the dialog box shown in Figure A-24.

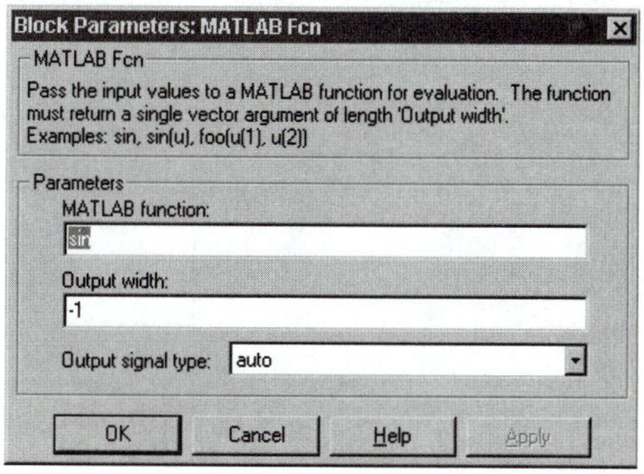

FIGURE A-24   Dialog box for the MATLAB Fcn block where
the user defines the file to be called by the simulation.

In the MATLAB function space, type in the name of the function file where the function was stored (here called mydemo). Output width refers to the size of the matrix that
is returned. In this case either 1 (since we return a scalar) or –1 (since the input and output are of the same width) will work. Generally make the more explicit choice, so 1 is
used. Click OK and run the simulation.

The results are shown below in Figure A-25.

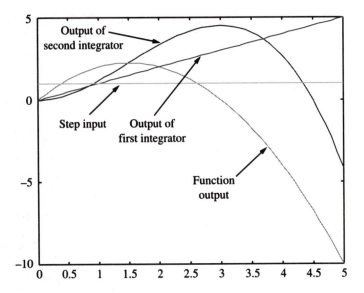

FIGURE A-25   Plot of all signals in the modified simulation versus time.

A.9.3   Using Multiple Inputs and Outputs

The previous example, although rather contrived, indicated the techniques required for incorporating function m-files in a Simulink simulation. Now a more physical example will be discussed. Consider the system showing in Figure A-26, consisting of a lumped mass, a spring, and a viscous damper.

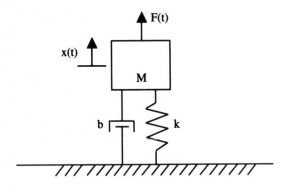

FIGURE A-26   Schematic diagram of a spring-mass-damper vibrating system.

The differential equation that describes the behavior of this system can be easily derived:

$$F(t) = m\ddot{x} + b\dot{x} + kx$$

where the overdot represents the derivative with respect to time.

A simulation of this system begins by solving for the highest derivative, as shown:

$$\ddot{x} = \frac{1}{m}\left(F(t) - b\dot{x} - kx\right)$$

For the purposes of this demonstration, we will be interested in finding the response of the system, $x(t)$, to arbitrary inputs, $F(t)$, as well as the total energy dissipated in the damping element. The rate of that energy dissipation (or power) is given as the product of the force in the damper and the velocity of the mass. Therefore:

$$\dot{E}_{diss} = P_{diss} = b\dot{x}^2$$

To find $x(t)$, therefore, the simulation will need to compute the second derivative of x as shown in the equation above and integrate it twice. The two integrators from our first example should come in handy. Beginning with that file, manipulate it so that it looks like Figure A-27.

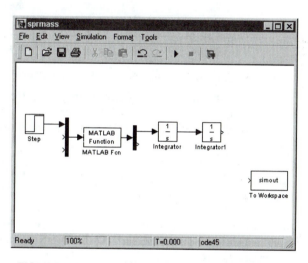

FIGURE A-27 Using Mux and DeMux blocks to handle multiple inputs and outputs for the function block.

Note that we've added two new blocks to the simulation, a DeMux block and a MATLAB Fcn block. The Fcn block we've seen previously, and the DeMux block, predictably, performs the inverse of the Mux block, taking a vector signal and splitting it into single signals. The first (top) DeMux output will be assumed to be the acceleration, and the second will be the rate of energy dissipated in the damper. The above equations make it clear that the function will need to have access to the input function—$F(t)$—as well as the velocity and the displacement of the mass. In addition, the power dissipated should be integrated to find total energy. Taking these issues into consideration, and adding one more Mux to store output signals in the workspace, the simulation can be further modified as shown in Figure A-28.

Note also that the labels on the blocks can be changed to reveal meaningful labels. By convention, we label the integrators with names indicative of the output, or result of the integration process.

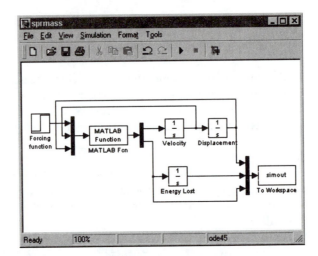

FIGURE A-28 Complete the simulation by connecting the blocks and adding an additional integrator to find total energy lost (file: `srpmass.mdl`).

At this point a discussion is in order regarding the use of loops in the simulation structure. Note that the velocity and displacement are going to be used to compute the acceleration of the system. This might seem a bit circuitous, like a snake swallowing its tail, but the fundamental structure of dynamic systems, as reflected in state space theory,[1] not only allows this, but requires it. The most important feature to notice in these loops is that there are integrator blocks within them. If a loop was formed that did not include some dynamic block, then the result would be an algebraic loop, which leads to numerical difficulties or may preclude the solution of the system. In general, the occurrence of algebraic loops (which are noted by MATLAB warning and error messages) indicates a poorly, or incorrectly formulated, simulation, and such loops should be regarded as errors.

Following the procedure already indicated for developing functions, we can write the function that will carry out the required computation for this simulation. The listing of this function is shown below:

pownacc.m

```
function out=pownacc(u)
%
%  Function used to compute acceleration
%  and power dissipated in a spring-mass-
%  damper system.
%
%  The inputs are assumed to be:
%  u(1) = F(t)
```

[1]Shearer, J. L., B. T. Kulakowski, and J. F. Gardner, *Dynamic Modeling and Control of Engineering Systems*, 2nd ed., Englewood Cliffs, N.J.: Prentice Hall, 1997.

```
%  u(2) = v(t)
%  u(3) = x(t)
%
%  The outputs are:
%
%  out(1) = acceleration
%  out(2) = power dissipated
%
%  Define the system parameters
%
m=1.0;
b=2.0;
k=3.0;
%
%  Compute the acceleration
%
out(1) = 1/m * (u(1) - b * u(2) - k * u(3));
%
%  Compute the power dissipated
%
out(2) = b * u(2)^2;
```

Several elements of this function set it apart from the previous examples. First, the input variable, u, is now a vector of several elements. This is indicated in the Simulink file by the fact that the input to the function block is a multiplexed line of three signals. The order in which they are connected to the multiplexor is the same as the order in which they appear in the function. In other words, the topmost signal (forcing function) is u(1), the second signal (connected to the velocity) is u(2), and the bottom signal (displacement) is u(3). The comments in the function listing are included as a reminder, but it is up to the user to ensure that the correct correspondence is recognized.

Second, the output is also a vector and the DeMux on the output behaves in a similar manner. Whatever is stored in out(1) in the function will be placed on the top, or first, signal from the DeMux, and so on.

Finally, we make use of many comments in the code. This is an essential part of any engineering model, and the importance of good comments cannot be overemphasized.

This file has been stored as pownacc.m. Let's return to the Simulink file, where the function block must be set up correctly. The filename should be indicated in the dialog box, and the output width should also be correctly set (see Figure A-29).

We will assume that all initial conditions are zero (the default for integrator blocks) and run the simulation. Once run, the results will be in the workspace and can be plotted:

```
» plot(tout,simout(:,1))
»
```

This leads to the plot shown in Figure A-30.

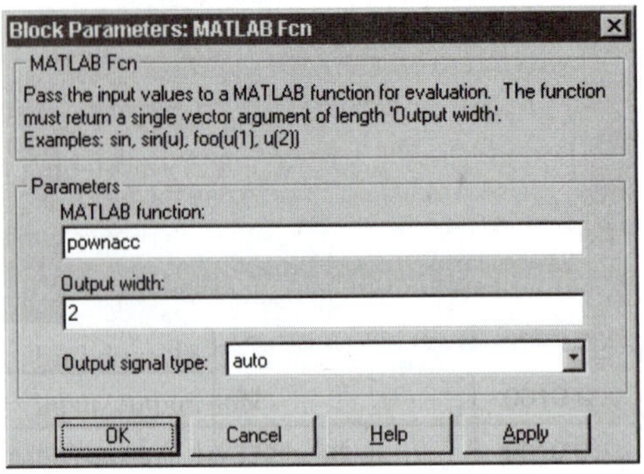

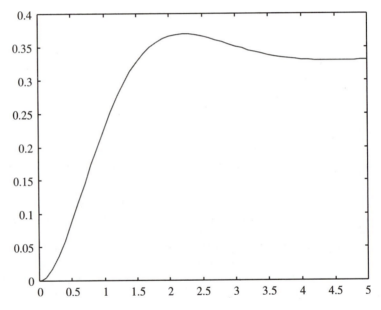

FIGURE A-29 Dialog box for the MATLAB Fcn block
when the function returns a vector.

FIGURE A-30 Plot of system response to a step input for the system shown in Figure A-26.

The parameters chosen (somewhat arbitrarily) in the function m-file lead to a slightly underdamped system. We can also look at power and energy dissipated by using the following plot command.

```
» plot(tout,simout(:,2),tout,simout(:,3))
»
```

From this we see the plot shown in Figure A-31.

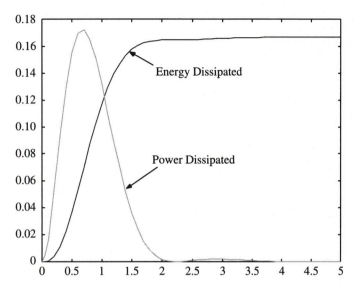

FIGURE A-31   Plot of power and energy dissipated
in the damper for the response to a step force input.

A.10   Concluding Remarks

This tutorial includes a minimal set of elements needed to carry out the simulations presented in this text. Both MATLAB and Simulink are exceptionally rich in their capabilities and features, and a simple tutorial cannot cover either program in reasonable depth. As with any complex engineering tool, there is no substitution for experience in using the software. The online help facility and the HTML-based helpdesk facility offer extensive documentation accessible from the workstation. The Mathworks web site (www.mathworks.com) is also an important resource with user-contributed files, questions, and answers.

INDEX